Micro-electronics:
the structure and operation of microprocessor-based systems

AF327478

Etymologia Scientifica

Electric (ĭle·ktrik). 1646.   [ad. mod.L. *electricus*, f. L. *electrum*, Gr. ἤλεκτρον amber; see -IC.]

   A. *adj.* 1. Possessing the property (first observed in amber) of developing electricity.   b. Charged with electricity.   2. Of the nature of, or pertaining to, electricity; producing, produced by, or operating by means of, electricity 1675.   3. *fig.* 1793.

   1. By Electrick bodies, I conceive..such as conveniently placed unto their objects attract all bodies palpable SIR T. BROWNE.   2. From e. fire (= *fluid*).. spirits may be kindled FRANKLIN.   3. The e. flash, that from the melting eye Darts the fond question COLERIDGE.

(Shorter Oxford English Dictionary on Historical Principles)

About our Author

Douglas Boniface's working life has evolved as a dichotomy, whose two periods were design and development of electronic instrumentation for the aircraft industry, and teaching in technical college and university. This has enabled him subsequently to transfer the practical experience, skills and knowledge gathered during those industrial years in to the minds and working capabilities of successive generations of students. It is this didactic quality which he now brings to *Microelectronics*, having taught GNVQ Microelectronics at advanced level on the pilot scheme for the syllabus on which the book is based.

His working life began on leaving school at the age of sixteen with an apprenticeship with W.E.Sykes Ltd., who made engineer gear cutting equipment. His graphic skills were considerable and he moved to become Design Draughtsman with E.N.V.Ltd., manufacturers of transmission equipment. That company moved away and he joined Page Engineering Company who made electromechanical and electronic equipment for the aircraft industry. He commenced as Design Draughtsman but, after holding a succession of posts, his ability and dedication were recognised and rewarded by promotion to Development Engineer involved with the design and development of aircraft equipment, and he worked on the electronic instrumentation for Concord, Harrier "Jump-jet", Tornado, the Lynx helicopter, and other types of military and civil aircraft.

During this industrial service he studied hard to advance future prospects, and gained the Higher National Certificate in Production Engineering (1965) and also the Higher National Certificate of Electrical and Electronic Engineering (1972).

In 1972 "Doug" Boniface made the important and courageous decision to change professions, one that would alter his whole life style. He decided it was time for him to qualify as a teacher and pass on the benefits of what industry had taught him. He obtained the Certificate in Education from London University in 1973, and later a Bachelor of Arts degree from the Open University In 1980,

His teaching career began in 1973 with the appointment as Lecturer at Highbury College in the Department of Marine Engineering, teaching engineering design, applied mechanics, and mathematics. After developing an interest in the growing technology of microprocessors he moved to the Department of Electrical and Electronic Engineering where he taught microprocessor software and hardware techniques, and also microelectronics and digital electronics. He taught GNVQ Advanced Engineering on a pilot scheme which is of particular relevance to this book since he has been involved with that qualification since it was initiated.

In recent years he moved to Fareham College as Lecturer, where he continues to teach GNVQ in preparation for students who look for careers as technician engineers. In addition to this assignment he also lectures in the Department of Electrical and Electronic Engineering at The University of Portsmouth. The university role makes him fully aware of the requirements needed by students who advance to higher study, which knowledge he communicates in this book.

Ellis Horwood, the Publisher at Albion Publishing Ltd
Chichester,   June 1966

MICRO-ELECTRONICS
the structure and operation of microprocessor-based systems

Douglas M. Boniface
Department of Electrical Engineering
Fareham College
and
Department of Electrical Engineering
University of Portsmouth

Albion Publishing

Chichester

First published in 1996 by
ALBION PUBLISHING LIMITED
International Publishers
Coll House, Westergate, Chichester, West Sussex, PO20 6QL England

COPYRIGHT NOTICE
All Rights Reserved. No part of this publication may be reproduced, stored in a retrieval system, or transmitted, in any form or by any means, electronic, mechanical, photocopying, recording, or otherwise, without the permission of Albion Publishing, International Publishers, Coll House, Westergate, Chichester, West Sussex, England

© Douglas M. Boniface, 1996

British Library Cataloguing in Publication Data
A catalogue record of this book is available from the British Library

ISBN 1-898563-32-2

Printed in Great Britain by Hartnolls, Bodmin, Cornwall

Table of Contents

Author's Preface

If the media is to be believed the world of microelectronics and computers is advancing rapidly and changing beyond all recognition. Yet, if we take a close look at the most widely used families of microprocessors, we find that they are based on techniques and architectures the fundamentals of which were developed nearly thirty years ago. What has changed is the range and complexity of the tasks being undertaken. Even at a domestic level many households now have either a computer or a games machine containing a microprocessor. They will probably also have another half dozen or so microprocessors, of which they are completely unaware, embedded in various pieces of equipment around the home.

This increasing role has led to the development of many new microprocessors. Often these new processors simply contain *more of the same* in less of the space as their predecessors and sometimes there is less of the same in even less of the space. However, several new architectures have evolved over the last few years. It has yet to be seen whether one of these will eventually dominate the market or whether this will be dominated by a completely new architecture, details of which have yet to be published. Whilst the future of microelectronics is assured its direction is less certain.

There are also significant changes occurring in the structure and financing of education. Thus in writing this book I have attempted to meet three criteria:

1   To convey basic concepts in such a way that they may be used either directly for one of the processors covered in the book or be easily adapted to meet the needs of other processors.
2   To meet the needs of new syllabuses such as GNVQ.
3   To cover the needs of these new syllabuses using the vast amount of equipment which already exists in schools and colleges

When studying microelectronics it is common practice for students to concentrate on one processor. If the first criterion to be met however it is important that from the start, and at each stage throughout their studies, students examine other processors and look for similarities. To avoid confusion caused by working with several processors the student is advised to develop solutions which are, as far as is practicable, independent of any processor. This 'processor independent' solution may then be translated to a solution for the processor of choice. Once a final solution has been established the student should then consider, in outline only, the translation to other processors. The worked examples and exercises in the book encourage this approach.

Whilst structured programming techniques have been common practice in the commercial world for many years, they have been largely ignored in the smaller engineering applications. The increasing complexity of applications has however necessitated their introduction and this has been reflected in more recent syllabuses. A chapter has therefore been devoted to the principles of such techniques and subsequent chapters use these techniques where appropriate.

Along with the new syllabuses have come changes to the methods of assessment. Typical of these are the new GNVQ syllabuses. To assist in the development of a portfolio of evidence most chapters contain exercises which are identified as *suggested for portfolio*. In the next section you will find a table indicating how these relate to each of the range statements in the current GNVQ Microelectronics (Advanced) syllabus. Also included are suggestions as to how the exercises might be adapted for the assessment of core skills.

Acknowledgements:
Finally thanks must go first to my wife Marie and my sons John and Peter for their assistance in the preparation of this book. Thanks must also go to my work colleagues for their *helpful* suggestions and to the innumerable students who have suffered during my attempts to find the best approach to various topics contained herein.

Good luck with your studies and I hope you find the world of microelectronics and microprocessors as fascinating as I do.

May 1996
D. M. Boniface
Fareham College, Hampshire

GNVQ Syllabus Cross Reference and Core Skills

BTEC Advanced GNVQ in Engineering
Unit 19: Microelectronic Systems (Advanced)   (issue dated November 1995)

The table below indicates the chapter or section in this book where each item in the syllabus range statements is covered. The final column suggests the exercises which might be used as evidence indicators to be included in the student's portfolio for the final award.

Element	Performance Criteria	Range	Covered in sections	Tested in Exercises
19.1	features	internal structure	2.2, 3.2, 3.3	Ex2.6
		internal operations	2.3, 3.3	Ex2.4
		bus connections	1.1, 4.9	Ex4.3
		bus functions	1.1, 2.3, 2.4	Ex2.6
		machine cycles	2.4	Ex2.6
		instruction sets	3.1, 6.., 7..	Ex3.1
	elements	microprocessor	1.1, 2.., 3..	Ex1.1, Ex3.1
		memory	1.1, 4..	Ex3.1 Ex4.2
		input/output	1.1, 4..	Ex4.3
		address decoding	4.9	Ex4.3
	factors	operating speeds	1.1, 3.3	Ex3.1
		address space	1.4, 4.9	Ex3.1, Ex4.3
		memory types	1.1, 4..	Ex4.5 Ex4.2
		bus widths	1.3	Ex4.2
		register widths	2.2	Ex2.4 Ex3.1
		instruction sets	3.1, 7.., 8.., 9..	Ex2.4 Ex3.1
		embedded control	1, 3.1	Ex3.1
		monitor routines	1.5	Ex1.5
19.2	problems	arithmetic	10.1	Ex10.6
		time delays	10.2	Ex10.3
		I/O routines	7.8, 7.9,	Ex7.6 Ex8.7 Ex10.3
	struct. design techniques	data flow charts	5..	Ex5.2
		structure charts	5..	Ex5.2
		top down	5..	Ex5.2 Ex5.6 Ex7.6 Ex8.7 Ex10.5

Elem-ent	Performance Criteria	Rang	Covered in sections	Tested in Exercises
	instructions	arithmetic	7.., 8.., 10	Ex 10.6
		logical	7.9	Ex 7.6 Ex 8.7 Ex10.3 Ex10.5
		data transfer	7.., 8..,9..,10.. 10.., 11..	Ex 7.6 Ex 10.3 Ex10.5
		program control	9..	Ex10.3 Ex10.5
	language	machine code	2.3	All programs
		assembler	2.3, 6	All programs
	check	functionality	5..	All programs
		errors	6.1	All programs
		logical design	5..	All programs
19.3	interface types	digital	11.3	Ex11.3 Ex11.5
		analogue	11.3	Ex11.4 Ex11.5
	protocols	timed	11.6	Ex11.1
		polled	11.2 11.4	Ex11.2 Ex11.4 Ex11.5
		interrupt driven	11.2 11.4, 11.5	Ex11.2 Ex11.4 Ex11.5
	interfacing devices	serial/parallel	10.5, 11.1 11.2	Ex11.1 Ex11.2
		parallel latch/buffer	4.11, App. F	Ex7.6 Ex8.7 Ex10.3 Ex10.5

Core Skills

Suggested below are the numbers of exercises in the text which might be used to provide some of the evidence for the achievement of core skills.

COMMUNICATION LEVEL 3

Elem't	Skill	Possible evidence indicators
3.2	Prepare written material	Ex3.1, Ex11.2, Ex11.4, Ex11.5
3.4	Read and respond to written material	Ex3.1, Ex4.2, Ex 4.3, Ex5.2

INFORMATION TECHNOLOGY LEVEL 3

Elem't	Skill	Possible evidence indicators
3.1	Set system options and setup, input data	Ex3.1, Ex5.2

APPLICATION OF NUMBER LEVEL 3

Elem't	Skill	Possible evidence indicators
3.2	Represent and tackle problems	Ex 10.2, Ex 10.5, Ex11.4

1

Microprocessor Based Systems: Introduction

1.1 MAIN FUNCTIONAL ELEMENTS

The term microcomputer means different things to different people. In engineering applications it might mean any thing from the controller in an automatic washing machine to the latest navigational aids in a luxury car or the heart of a fully automated production process. In the commercial world it might mean anything from a simple word processor to a powerful information retrieval and accounting system. Whatever its size and application, the most common configuration of a microcomputer consists of four basic elements interconnected as shown in Figure 1.1 (even if they are sometimes in the same chip).

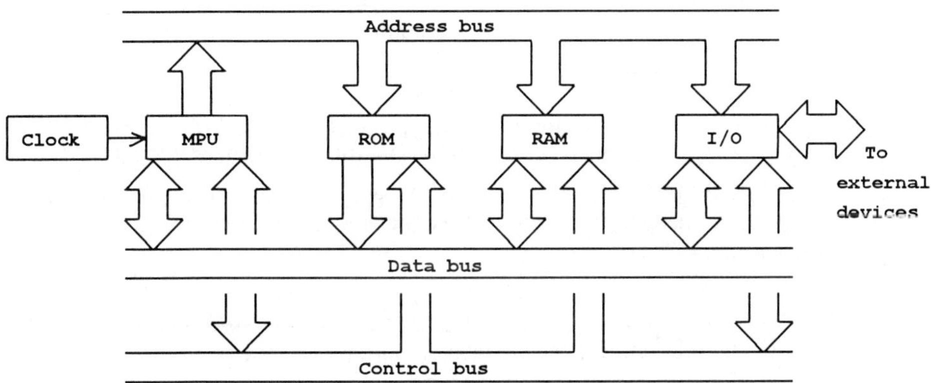

Figure 1.1

MPU stands for **MicroProcessor Unit**. The microprocessor is the clever part of the system and is responsible for executing a wide variety of data transfer, arithmetic and logical instructions. Although MPU is regarded as the clever part of the system, it is important to realize that it can do very little without being given precise instructions telling it what is required and how it is to be achieved. These instructions are held in memory of which there are two main types: *ROM* and *RAM*

ROM stands for **Read Only Memory**. This type of memory has programs and data entered at the time of manufacture. Its contents are permanent and cannot be changed. ROM will always contain some instructions indicating what is to happen when the system is first switched on and, in smaller systems it may contain the entire set of instructions for operating the equipment. In larger systems other programs may be loaded from external devices such as disk drives and ROM will often contain many utilities for use by these programs.

RAM stands for **Random Access Memory** but is more accurately described as **read - write memory** since its contents may be changed. In small systems, RAM may be used merely to hold data from the process it is controlling. In larger systems, however, this type of memory may hold both data and programs loaded from an external source.

Both ROM and RAM may be thought of as consisting of a long column of pigeon holes known as **memory locations** or **storage locations**. Each memory location will contain a binary number that may be part of an instruction or may be an item of data.

I/O stands for **Input/Output** and it is through these devices that the microprocessor is able to communicate with the outside world. In a typical control application I/O devices will be used to connect the computer to such items as relays, switches, sensors etc.. In commercial applications they are used to connect to keyboards, VDUs, printers etc..

Address Bus. If RAM and ROM are to be thought of as a long column of pigeon holes or memory locations then, to be able to identify a particular memory location, it is given a number known as its **address**. When the microprocessor needs access to a memory location it puts the address of that location onto the **address bus**.

Data Bus. Once the microprocessor has identified a memory location, data can pass between the microprocessor and memory along the **data bus**.

Control Bus. Although most parts of the control bus are not dealt with in detail on this course you should realize that, for the system to operate effectively, signals are needed to control a number of aspects of the system such as direction of data flow, restricting the time when data is available etc. These signals are passed around the system on the **control bus.**

Clock Almost every microprocessor based system has an oscillator which generates a rectangular wave with a frequency which may be from a few hundreds of kilohertz to a few hundreds of Megahertz. It is known as the **system clock** for historical reasons. In most systems, but by no means all, the rising and/or falling edges of the clock signal is used to initiate each step in the execution of instructions. When data sheets refer to the **operating speed** of a processor they are usually refering to the maximum speed of the clock signal.

1.2 BITS, BYTES and NIBBLES
(If you are uncertain about binary numbers and the relationship between binary, decimal and hexadecimal, refer to Appendix G before reading this section.)

Some types of early computer (analogue computers) tried to represent the values in a problem with an equivalent voltage. This presented all sorts of problems with scaling errors and the need for equipment which could detect and respond to small changes in voltage. It is far easier to determine if a particular voltage is high or low than to determine its precise value.

The terms high and low are relative terms. In digital electronics they are relative to the voltage at which the main computer integrated circuits (ICs) are operating. Many computers operate at 5V and so any signal near 0V is regarded as a low voltage and any signal near 5V is regarded as a high voltage. This leaves a band in the middle which is undefined and it is the manufacturers job to ensure that signals in this range never occur.

Because we are no longer interested in the precise value, we can use any symbol we like for high and low and there are good mathematical reasons for choosing the binary digits 0 and 1. (The term **BIT** is used to indicate a **BI**nary digi**T**.) Suppose I wanted to represent the decimal value 6 in an electrical format:

The binary equivalent of decimal 6 is 110 (written $6_{10} = 110_2$). I can now represent 6_{10} using three wires e.g.

i.e. high, high, low or HHL

In a similar way I could represent 26_{10} (11010_2) with five wires set to HHLHL.

The first microcomputers to be in really wide general use had MPUs with a data bus which consisted of eight wires or tracks. A name was needed for a package of data on the data bus and the term **BYTE** was chosen to mean a group of 8 bits. When a low value number is to be sent along an 8 bit data bus leading zeros are added,

e.g. $6_{10} = 0000\ 0110_2$ and $26_{10} = 0001\ 1010_2$

Long binary numbers are difficult read. Splitting a byte into two, four bit groups (**NIBBLES**) and expressing them in hexadecimal means that quite large binary numbers can be easily represented. Using fewer characters reduces the chance of error and has obvious advantages for text books and technical manuals.

1.3. DATA SIZE

MPUs with an 8 bit data bus are known as 8 bit processors; those with a 16 bit data bus are known as 16 bit processors etc.. Although today there are many processors with 16 and 32 bit data buses, many of the support ICs were developed for use with 8 bit devices and are still widely used. For this and other reasons, most processors have instructions which still allow data to be accessed 8 bits at a time.

1.4 MEMORY SPACE

The number of lines in the address bus indicates the maximum number of memory locations that may be addressed directly by the processor. If a processor existed with a ten bit address bus it could address 2^{10} = 1024 memory locations and, since this is near to one thousand, 1024 memory locations each capable of holding 8 bits of data became known as 1 kilobyte. Many early processors had a 16 bit address bus and so were capable of addressing

$$2^{16} = 65536 \text{ memory}$$
$$\text{but} \quad 2^{16} = 64 \times 1024$$

and so this quantity is often referred to as 64 kilobytes or 64K. For similar reasons a processor with a 20 bit address bus can address 2^{20} = 1048576 locations and, since this quantity is close to one million, it became known as 1 megabyte.

The size of the address bus does not place an absolute limit on the memory size. In some systems, banks of memory are switched in and out to give the effect of a much larger memory space. On the other hand, in the modern PC, it is quite common to find that the system has much less memory than the processor is capable of addressing and techniques have evolved which use disk space for a memory image.

1.5 OPERATING SYSTEM AND SYSTEM MONITOR

The way in which a microprocessor responds when the system is first switched on varies from one type of processor to another. In general, however, one of two techniques are used. Either the processor goes to a particular memory location and starts executing program from there, or it goes to a particular location to read the address of its first instruction. In either case ROM is needed for the initial procedures. What happens next will depend on the functions required rather than the type of processor being used.

If the processor is being used for a totally dedicated and self-contained control application the program in ROM will probably start by setting up a few I/O devices and then proceed directly to execute the main program.

In more versatile control applications and for a typical desk top computer there will be a need to perform tasks such as: select the process required, load a program from an external source and run it etc. These processes require two set of programs, those required to perform the main tasks and those to provide a convenient user interface. Programs for setting up I/O devices and providing a user interface are often grouped together under the heading of System Monitor or Operating System. Operating systems frequently contain routines which may be used directly by the programmer e.g. reading the key board.

The size and location of the operating system also varies with the application. Some applications will have the entire operating system in a few kilobytes of ROM, others will use megabytes of ROM. A typical PC will have some of its operating system in ROM and some on disk. In this case the ROM based part will usually do a few basic operations such as system checks and setting up some I/O devices before loading the main part of the operating system from disk into RAM. The advantage of this approach is that it provides an easy and cheap way of changing operating systems or upgrading the current one.

Exercise 1.1   *** *suggested for portfolio* ***
Ideally students should have access to a small single board computer for the following exercise. If this is not available they should make use of the figure 1.2 which shows the layout of the main integrated circuits on an educational system.

Examine a small single board computer and make a note of the numbers on each of the ICs. Then use catalogues and manufacturers data sheets to answer the following questions:

  a)   Identify the function of each component.
  b)   Refer back to the block diagram of a micro computer given on page 1
       and state the ICs which have been used for each of the four main blocks.
  c)   How many address lines and data lines does this processor have?
  d)   How many memory locations can be addressed by this processor?
  e)   How much memory (ROM and RAM) has been provided on the system?

Exercise 1.2
Refer to a data sheet for the microprocessor on the single board computer in exercise 1.1 and establish the supply voltage and maximum clock frequency. Using suitable instruments measure the voltage supply to the microprocessor (VCC and GND) and the actual frequency of the clock signal which arrives at the clock pin (usually shown as either CLK or ϕ). Account for any differences between the figures on the data sheet and those determined experimentally.

Exercise 1.3   *** *suggested for portfolio* ***
Produce a diagram showing the layout of the microprocessor and the main components used for memory and I/O on the board refered to in exercise 1.1. On each component mark the position of the pins used for data lines and address lines. Then show how the tracks on the PCB which form the address and data busses connect these components. (If you did not have a physical board to deal with and used Figure 1.2 for earlier work you should draw what you believe to be sensible routes for the busses.)

Exercise 1.4   **** suggested for portfolio ****

Using catalogues from several large suppliers of electronic components make a list of 8-bit, 16-bit and 32-bit microprocessors together with their price. Using the price range you have found draw up a table classifying the processors as either low cost, medium cost or high cost. (For the purpose of this exercise you should treat microprocessors and microcontrollers as the same.)

Exercise 1.5   **** suggested for portfolio ****

Imagine that you have been asked to write some software which uses a desk top PC for a control application. Make a list of the operating system routines which you think might be useful.

Component list for figure 1.2

U1	Z80 - CPU	U6	2516	U11	Z80 - CTC
U2	74HCT74	U7	2516 or 6116	U12	75491
U3	74LS14	U8	6116	U13	75492
U4	74LS90	U9	74LS139	U14	8255 PPI
U5	74LS139	U10	Z80 - PIO	U15	75492

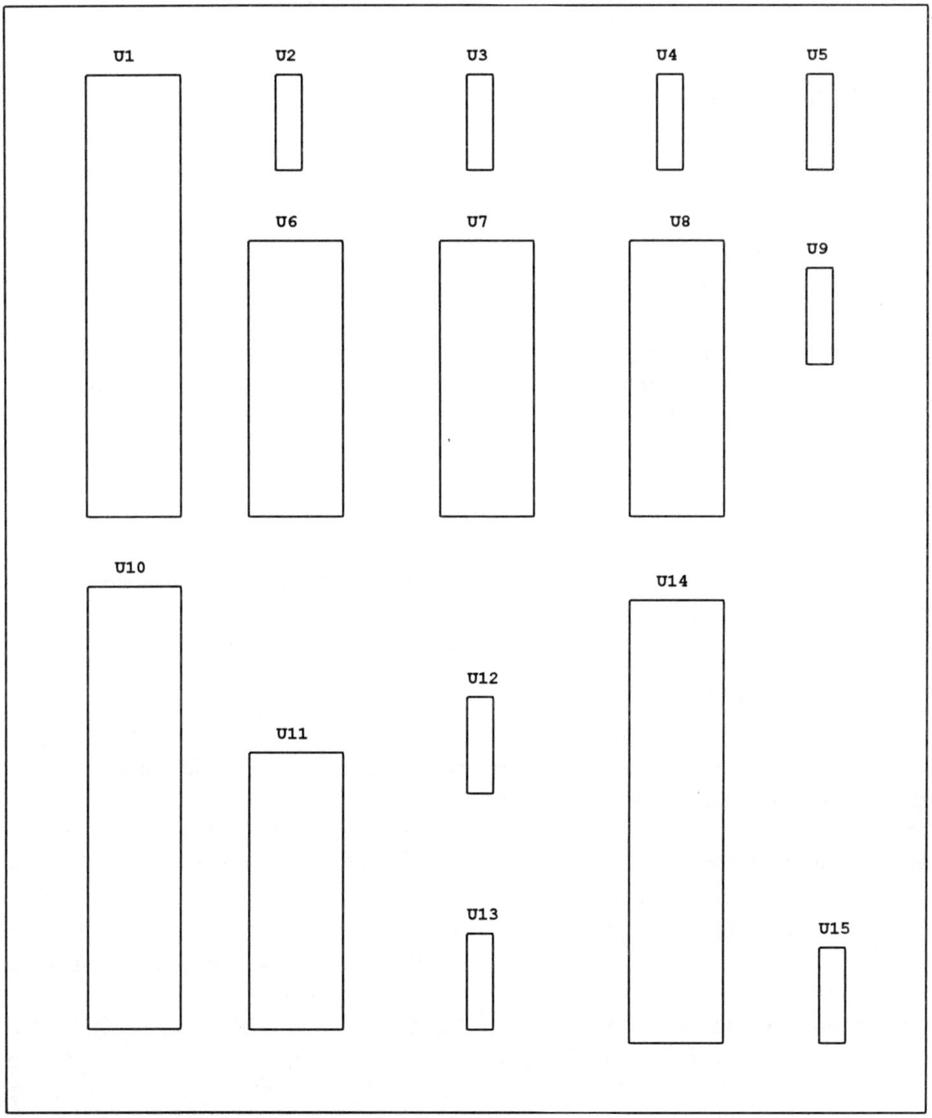

U1 U2 U3 U4 U5

U6 U7 U8 U9

U10 U14

U11 U12

U13 U15

Figure 1.2

2

Microprocessors - their operation

2.1 ESSENTIAL PARTS

Although the internal details of a microprocessor vary considerably from one processor to another they almost all have the following elements in common:

(a)   Somewhere to hold the details of the instruction on which it is currently working.

(b)   Somewhere to hold details of the location of the next instruction.

(c)   Internal registers for the storage of data and/or addresses.

(d)   A unit for performing arithmetic and logical operations.

(e)   Somewhere to store information about the nature of the results of the last operation.

N.B.   In the notes and exercises which follow the reader should remember that the diagrams of the internal structure of the processors have been simplified to aid clarity. The diagrams are sufficiently good models to meet the needs of this and many other courses, but if the reader wishes to work in depth he should obtain appropriate data sheets from the manufacturer.

2.2 MAIN FEATURES OF A MICROPROCESSOR

Figure 2.1 shows the author's model for the internal structure of a Z80 microprocessor. You should read through the descriptions given for each of its parts once and then return to these descriptions when necessary.

Instruction Register This is where the operator part of an instruction is held whilst the instruction is being executed.

Instruction Decoder Once the operator part of an instruction has been placed in the instruction register, this unit is responsible for working out what needs to be done.

Control and Timing In order to execute an instruction, it will be necessary for the processor to send control signals to other units inside the processor and to other parts of the system. From time to time it will also be necessary for the processor to respond to control signals coming from external sources.

Acc   Acc is short for **accumulator** and it is an 8 bit register. When an arithmetic or logical operation is to be carried out, one of the operands is usually placed in the accumulator.

ALU This stand for Arithmetic and Logic Unit. The ALU contains the circuitry necessary to perform arithmetic and logical operations. The result of these operations is usually written back to the accumulator.

Flags After an ALU operation it is often necessary to be able to answer questions such as "Was the result zero?", "Was the result too big to fit in the accumulator?" etc. This information is held in the flags register.

PC In this context PC stands for **Program Counter** and it is a 16 bit register. For the majority of the time the program counter holds the address of the next byte of program to be accessed.

MAR This stands for **Memory Address Register** and it is a 16 bit register. It is often used when the processor needs to access a memory location which is out of sequence from the program.

B, C, D, E, H, L These are 8 bit general purpose registers which may be used in pairs (BC, DE and HL) to act as 16 bit registers. They may be used for storing data and addresses for use by the processor.

Multiplexers There are a number of different registers that might need access to the address bus and a number that might need access to the data bus. The multiplexers respond to signals from the control and timing unit and ensure that the correct data pathway is set up. (On many diagrams the term multiplexer is abbreviated to **MUX**.)

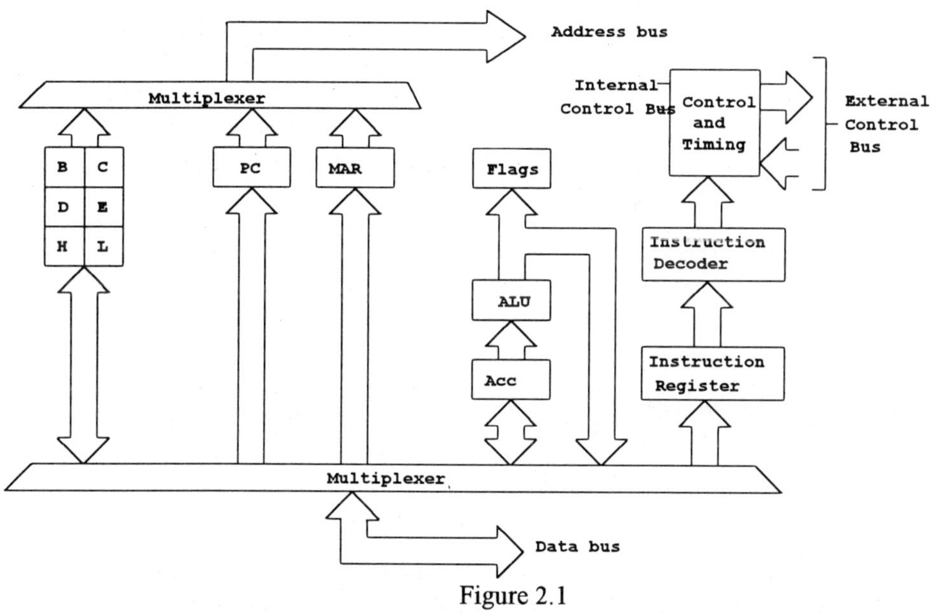

Figure 2.1

The connections to the address and data buses are made *via* tristate buffers (not shown) as are many of the lines in the control bus. This is necessary since, in some systems, it will occasionally be necessary to allow other devices to use the buses. The tristate-state buffers allow us to electrically isolate the processor from the rest of the system whilst leaving them physically connected.

In connection with the buses it should be noted that the address bus is unidirectional (addresses are only sent out on this bus), the data bus is bi-directional (data can flow in either direction) and the control bus is a collection of unidirectional signals (some are input only and some output only).

Now complete Exercise 2.1 at the end of this chapter.

2.3 FETCH-EXECUTE CYCLE

If effective use is to be made of microprocessor instructions some understanding of how they are executed is essential. There are two parts to an instruction; the operator and the operands. The operator part indicates what is to be done and the operands are the things on which the operator operates.

A typical line from the listing of a program written for the Z80 might be:

```
1820   3a 4c 19              ld      a,(194c)
```

The interpretation of this line is:

```
1820
```
This tells us that the instruction starts at address 1820.

```
3a 4c 19
```
This is the machine code form of the instruction and, when translated into binary, is the only form of the instruction that can be understood by the processor.

```
ld   a,(194c)
```
This is often referred to as the assembler or mnemonic form of the instruction. It is this form that is used by a programmer when he first writes his program. When the program is written it must be converted into machine code for use by the processor.

In the mnemonic form a Z80 instruction has the general structure:

        operator, destination, source

i.e.

operator	ld	load or copy information.
destination	a	the information is to be placed in the accumulator.
source	(194c)	the information is currently in address 194c.

In this case the operator is clearly ld (load), the source operand is the contents of memory location 194c and the destination operand is the accumulator.

When a processor executes an instruction it goes through the following sequence:

1.   Fetch operator.
2.   Fetch operand.
3.   Fetch more operands if needed.
4.   Operate on operands (execute the instruction).

(5.   Go back to 1 and repeat for the next instruction.)

We can illustrate this process with the instruction discussed above.

If the contents of memory location 194c were a6 then, at the start, we would have:

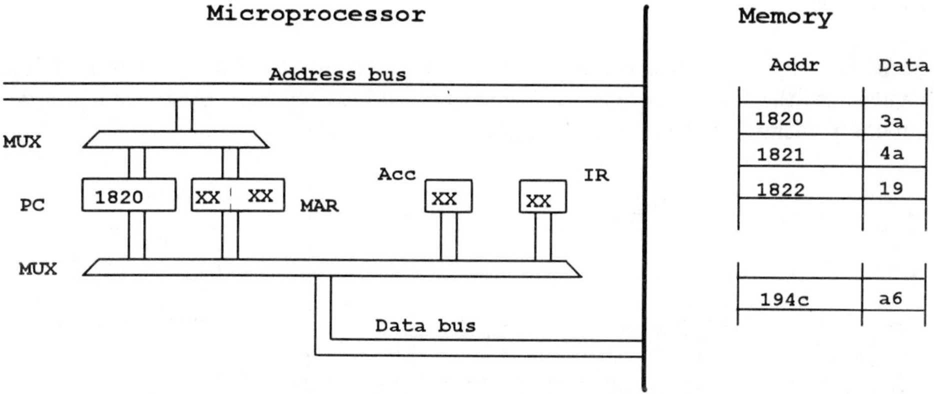

Step 1. Program counter is put onto the address bus and the data at address 1820 is copied into the instruction register. As soon as the program counter has been used it is incremented and so at the end of step 1 we have:

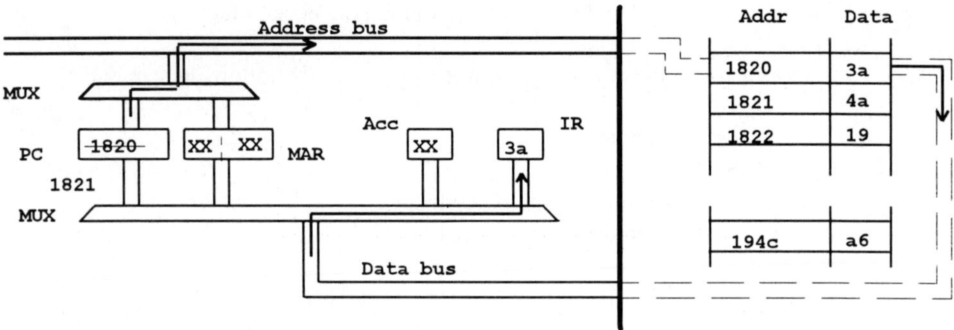

The instruction 3a is decoded and the processor now knows that it needs a 16-bit address. The data bus is only 8-bits wide and so it is necessary to fetch the address in two stages.

Step 2. Program counter is put onto the address bus and the data at address 1821 is copied into the low order byte of the memory address register. Again the program counter is incremented and so at the end of Step 2 we have:

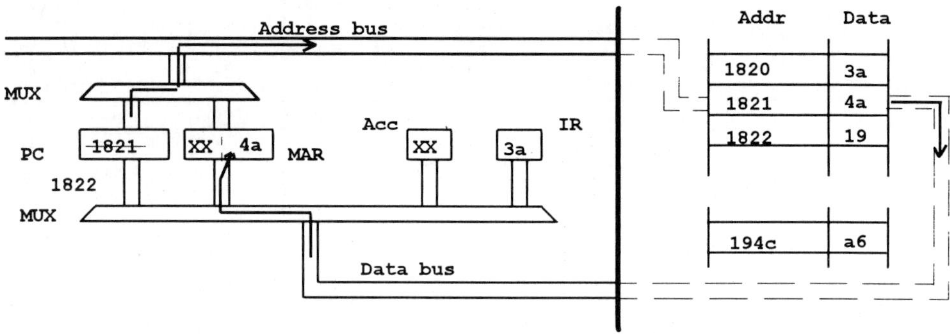

Step 3. To get the second byte of the address the program counter is put onto the address bus and the data at address 1822 is copied into the high order byte of the memory address register. Again the program counter is incremented and so at the end of Step 3 we have:

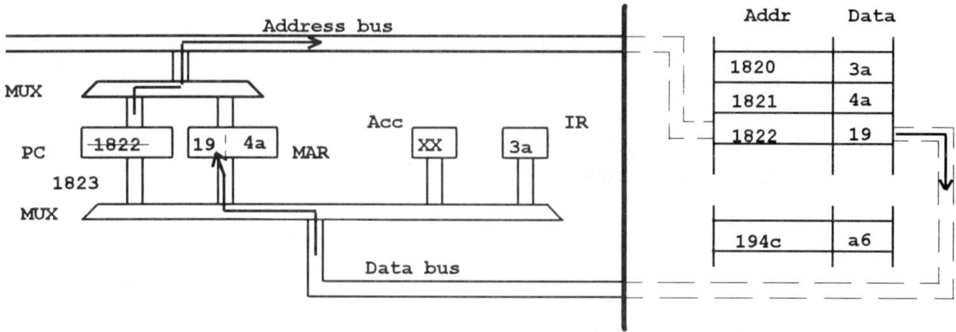

Step 4. The processor now has the complete address and so this time it is the memory address register which is put onto the address bus and the data at address 194a are transferred to the accumulator. The program counter has not been used and so it is not incremented. Thus after Step 4 we have:

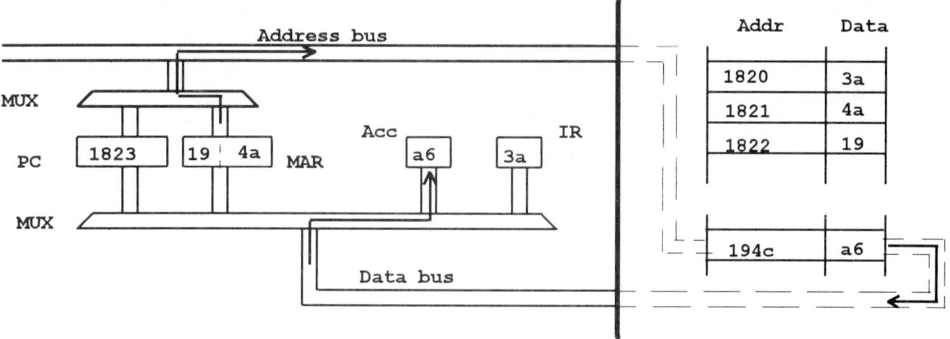

Note that at the end of Step 4 the program counter is pointing at the next address in memory. This is where the next instruction starts and so the next step would be to put the program counter onto the address bus and get the operator part of the next instruction.

The steps above may be expressed in a text format as follows:

Step 1   PC -> Address bus
data at 1820 -> IR
increment PC

Step 2   PC -> Address bus
data at 1821 -> MAR low
increment PC

Step 3   PC -> Address bus
data at 1822 -> MAR high
increment PC

Step 4   MAR -> Address bus
data at 194c -> accumulator

Now complete exercise 2.2 at the end of this chapter.

2.4 MACHINE CYCLES AND TIMING DIAGRAMS

To ensure that data flows in the right direction, most processors distinguish between operations which **read from memory** and operations which **write to memory**. This is achieved with a signal or signals in the control bus and the processes are referred to as read cycles and write cycles. Processors such as the Z80 and 80x86 also distinguish between memory and I/O data transfers Each part of the process is initiated by either the rising edge or the falling edge of a clock pulse. On a Z80 the relevant control signals are:

    RD   Read - Normally high but goes low when performing a read operation
    WR   Write - Normally high but goes low when performing a write operation
    MREQ   Memory request - goes low when the processor needs to access memory
    IORQ   I/O request - goes low when the processor needs to access an I/O device

The relationship between these signals can be seen in the following timing diagrams:

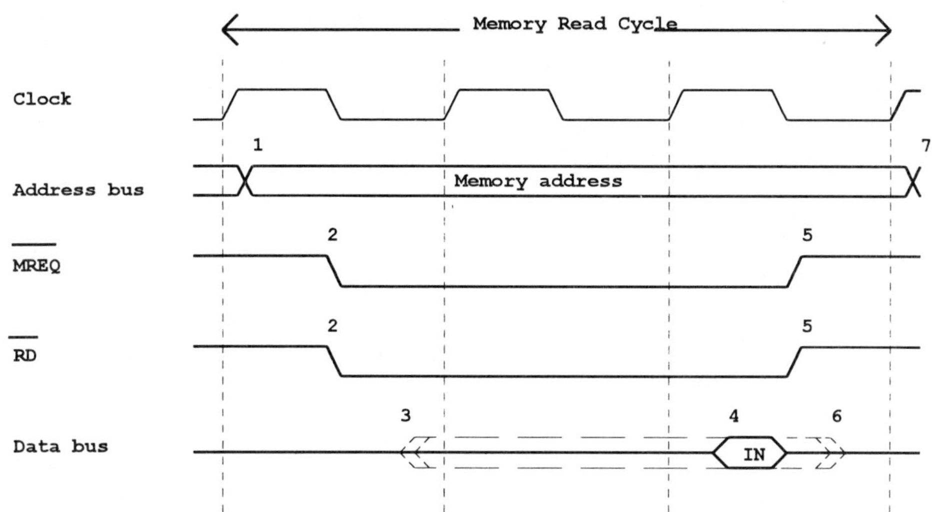

Figure 2.2

Key

1   The rising edge of the first clock pulse signals the processor to put the required address onto the address bus.

2   Once the address has had time to stabilize MREQ is taken low to signal to other components in the system that the address bus now holds a valid memory address. RD is also taken low indicating the direction of data flow.

3   Some time between 2 and 4, depending on the speed of the components, data will start to appear on the data bus.

4   This is the time when the processor reads the data that has appeared on the data bus.

5   Once the data has been taken in the processor deactivates MREQ and RD. This has the effect of causing the memory device to shut down and the data on the data bus to start collapsing.

6   Some time after 5 the data on the data bus, held there by stray capacitance, will have finally collapsed.

7   Start of new operation

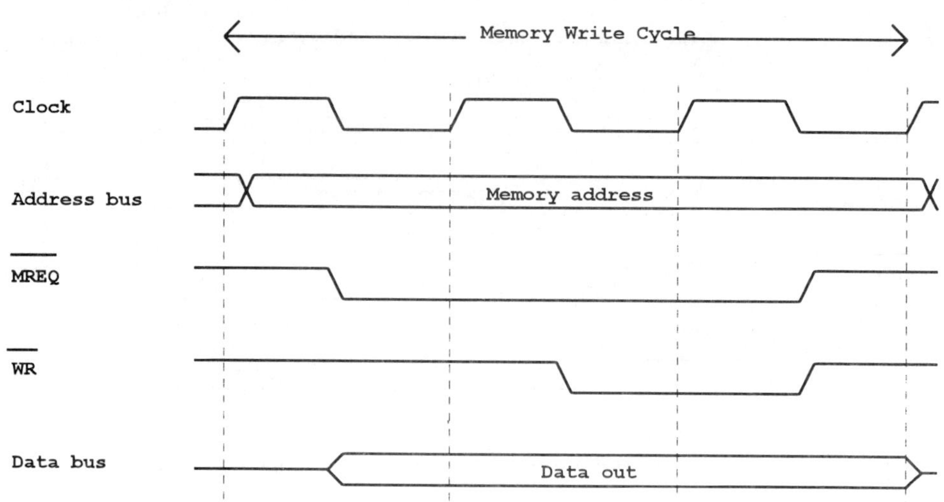

Figure 2.3

The **memory write cycle** is similar to the **read cycle**. Notice however that this time it is **MREQ and data** which are sent out together. Once the data has had time to settle on the data bus and over come stray capacitance the WR signal goes low, indicating that the data bus holds valid data. Also notice that WR is taken high before any attempt is made to remove the data. This ensures that the memory device does not take in collapsing data.

The Z80 also distinguishes between reading the operator part of an instruction (op-code fetch) and any other memory read operation by means of another control signal designated M1. The timing diagram for an op-code fetch is given below.

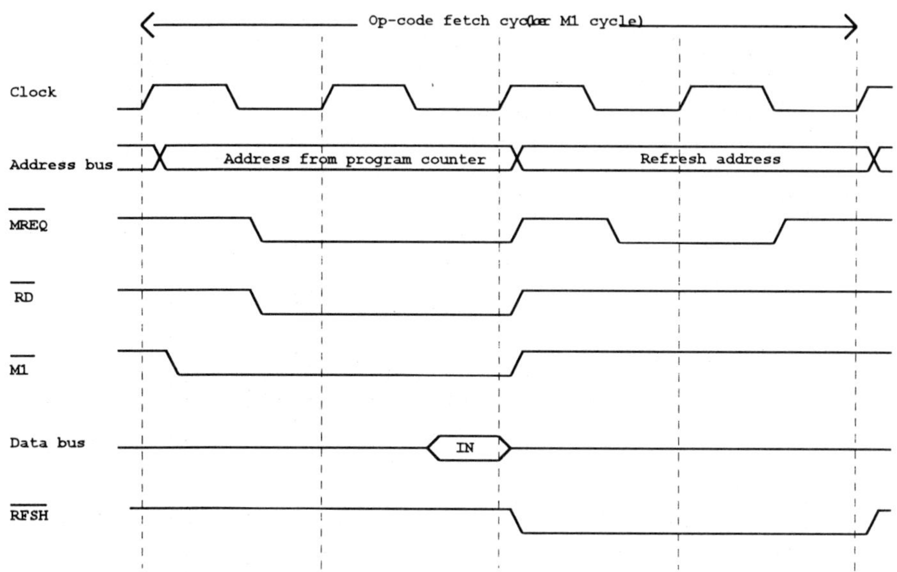

Figure 2.4

(The purpose of the refresh address and the control signal RFSH will be dealt with later.)

Now complete Exercises 2.3, 2.4 and 2.5 at the end of this chapter.

The 8086 processor is designed so that it may be used with other processors and some of its control signals may change their function. When used in a single processor system, the 8086 has similar control signals to the Z80 except that MREQ and IORQ are combined into a single pin M/IO which is high for memory access and low for I/O access. It is normal practice to combine the signals M/IO, RD and WR with some external logic to generate four new control lines designated MEMR (memory read), MEMW (memory write), IOR (I/O read) and IOW (I/O write). These have timing diagrams similar to those for the Z80 e.g.

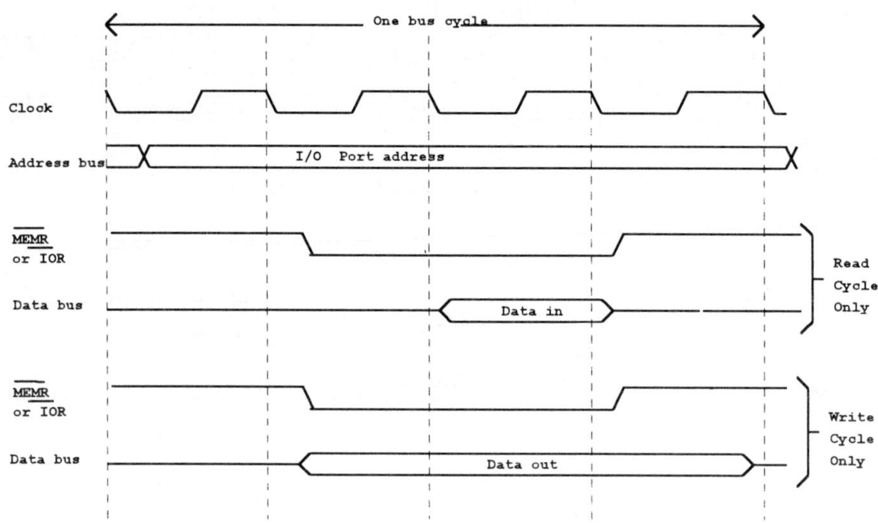

Figure 2.5

In all the timing diagrams encountered so far, events have been synchronized by the clock pulse. Either a rising edge or a falling edge of the clock pulse is used to initiate an event. The bus cycle for the 68000 is **semi-synchronous** in that it starts by using clock edges to initiate events, but then waits for an external data transfer acknowledge signal (DTACK) before continuing. The DTACK signal is usually generated by some sort of time delay circuit which has been designed to allow adequate time for the memory device to respond and the data to become stable. A timing diagram for a memory read cycle is shown in Figure 2.6.

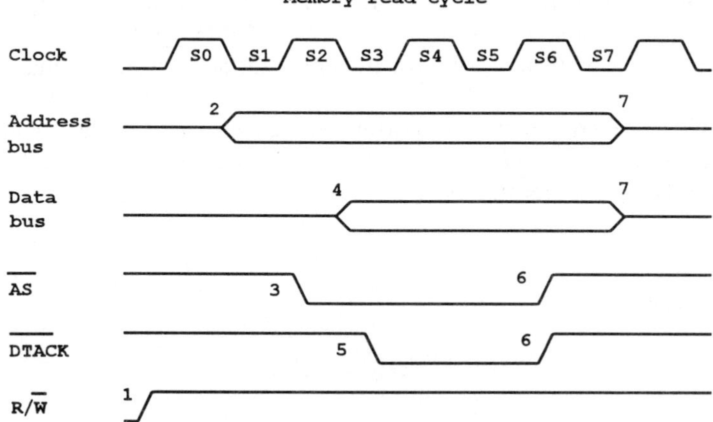

Figure 2.6

Key

1   Read/Write line is set to read.
2   The address is put onto the address bus
3   The address strobe control line (AS) is activated to indicate that the address bus holds a valid address.
4   The memory device puts its data onto the data bus.
5   Supporting circuitry takes DTACK low and this is used to latch the data into the processor. If, by the rising edge of S4, DTACK has not gone low, the processor inserts an extra clock pulse between S4 and S5 before looking at DTACK again.
6   Having found DTACK low, the processor responds by taking AS high indicating that the data has been read. The supporting circuitry responds by taking DTACK high.
7   The data on the data bus collapses and the processor removes the address from the address bus.

A memory write cycle follows similar principles. This time, however, DTACK is used to ensure adequate time has been allowed for data to be written to memory and it will be between S6 and S7 that extra clock pulses will be added if necessary. The timing diagram for a memory write cycle is shown in figure 2.7.

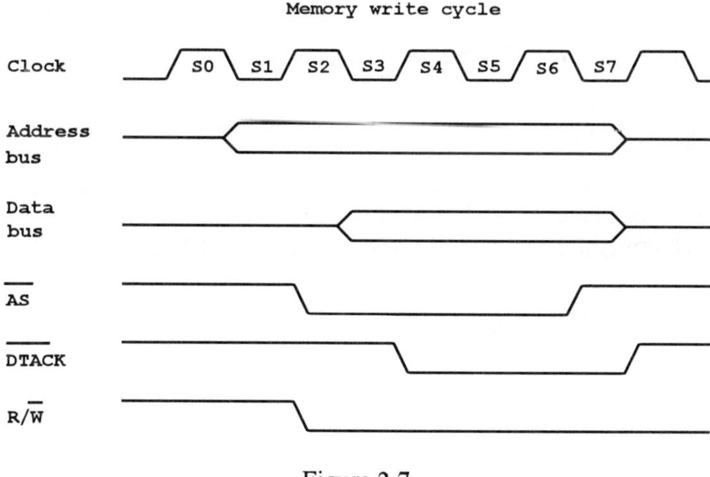

Figure 2.7

Exercise 2.1
Appendix C and Appendix D contain details of the 8086 and 6800 processors. Examine these processor details together with Figure 2.1 and determine which parts correspond to the items (a) to (e) listed in Section 2.1.

Exercise 2.2
The line of code

```
1897    32 34 1c              ld      (1c34),a
```

means copy the contents of the accumulator into memory location 1c34.

    (a)   Identify the source and destination operands.

    (b)   If the program counter is at 1987 and the accumulator contains 55, express the sequence of events that occur in the format given at the end of section 2.3

Exercise 2.3
In section 2.3 we looked at the fetch-execute cycle for the instruction `ld a, (194c)`. With reference to this instruction answer the following questions:

    (a)   How many op-code fetch cycles, memory read cycles and memory write cycles are used?

    (b)   How many clock pulses does it take?

    (c)   Draw a timing diagram for the entire instruction. You should show the address bus, M1, MREQ, RD, WR and the data bus.

Exercise 2.4   *** *suggested for portfolio* ***
Repeat exercise 2.3 for the instruction `ld    (1c34),a`.

Exercise 2.5
The timing diagram for the input read and output write cycles of a Z80 (reading from and writing to I/O devices) is shown in Figure 2.8. Write a key for this diagram similar to that written for the memory read cycle (Figure 2.2).

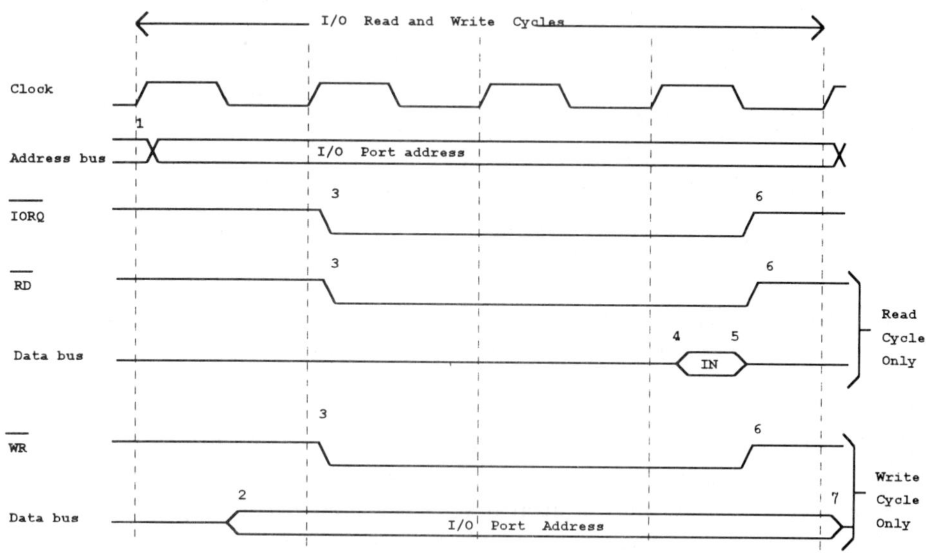

Figure 2.8

Exercise 2.6   *** *suggested for portfolio* ***

Draw a complete timing diagram for the following instructions:

a)	ld	a,(1234h)	(Z80)
b)	mov	ah,(1234h)	(8086)
c)	move.b	$1234,a4	(68000)

3

Microprocessor Types

In this chapter we shall be looking at a range of different types of microprocessor and the factors which might influence the choice of a microprocessor for a particular task.

3.1 MICROPROCESSORS VERSUS MICROCONTROLLERS

At one time the terms microprocessor and microcontroller represented two significantly different products. There are still many devices which are clearly microprocessors and many which are clearly microcontrollers but, as the complexity of each has grown, there are many which are difficult to classify. As a first attempt we might say that a microprocessor is a device which has been designed primarily for computing and a microcontroller is a device designed primarily for control applications. Using these definitions, then most of the 80x86 family of processors are clearly a microprocessors, and devices such as the 8051 are clearly microcontrollers. The 80x86 family of processors, particularly the 80186, have however been used for control applications and it is quite possible to build a computer using an 8051. In the middle we have processors such as the Z80 which, in the late 70's through to the mid 80's was widely used for small computers. The Z80 is still quite widely used but now, almost exclusively, for control applications. A better distinction can be drawn if we look at the needs of a computer compared to a controller.

The most common needs of a computer are that it can manipulate, tabulate and store large quantities of data and that it can perform calculations. Devices classed as microprocessors usually have a good selection of arithmetic instructions such as add, subtract, multiply and divide They also have instructions which make it easy to move, manipulate and search large blocks of data.

A controller must be able to respond to alarm signals, inputs from sensors and control external devices. These activities often mean dealing with individual bits within a byte. It is not uncommon to find that control programs are quite small and need to deal with only a few tens of bytes of data. Devices classed as microcontrollers will often have RAM,   ROM and I/O on the same chip as the microprocessor, and will have a good range of bit test and manipulation instructions. They tend to have a versatile interrupt structure (see Chapter 11) which makes it easier to respond to alarm signals.

The complete list of instructions available for each processor is called its *instruction set* and this is usually provided in the data sheet for the processor. Appendix E gives an outline of the range of instructions for a number of processors.

3.2 PROCESSOR ARCHITECTURES

The processors dealt with in earlier chapters have what is sometimes referred to as a **Von Neumann architecture**. This is by far the most common architecture and, roughly translated, it means that the processor has one address bus and one data bus. This causes a bottle neck since the busses can only do one thing at a time.

The 80x86 processors attempt to overcome this problem with an instruction queue. An instruction is fetched whenever there is space in the queue and the busses are not being used for anything else. Once an instruction has been decoded there may be a need to get data from somewhere else. In this case the *instruction fetch* operations are suspended whilst the data is fetched. There will be times, such as when a jump instruction is encountered, when it will be necessary to dump the entire queue and start again. Not withstanding these restrictions the instruction queue does lead to a saving in time.

The need to suspend *instruction fetch* whilst data is being retrieved or stored could be avoided if data and code were in different memories. Processors with a **Harvard architecture** (see Figure 3.1) achieve this by having separate address and data busses for code and data .

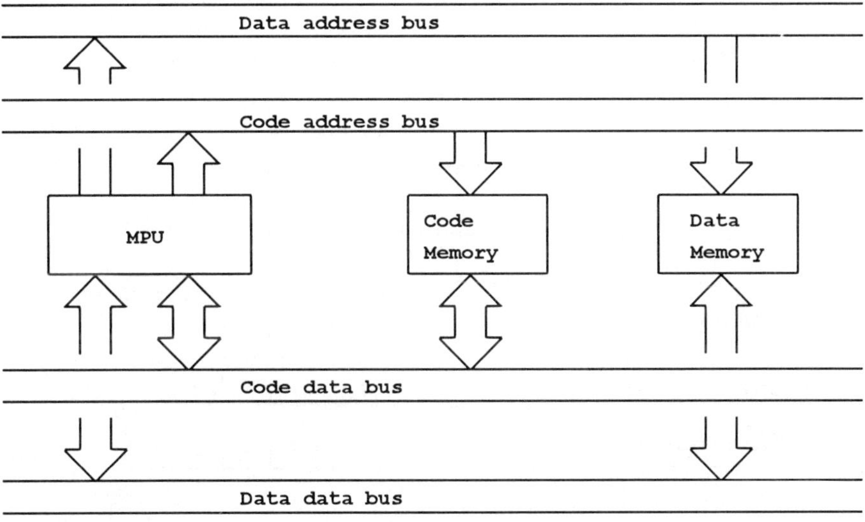

Figure 3.1

3.3 CISC and RISC

CISC stands for Complex Instruction Set Computer. The range of instructions built into early processors was relatively small, and complex operations needed to be constructed from a number of simpler operations. Multiply operations, for example, were often implemented in software as a repeated addition. As the variety of processor applications grew, so did the requirement for more complex operations. After a while it became apparent that, not only were the complex operations taking a lot of processor time, but the time taken to write the corresponding code was taking an unacceptably large amount of the programmer's time. This led to the development of the CISC processors of which the 8086 and 8088 were probably the first to go into really wide general use.

Ever since their introduction, the use of CISC processors has continued to expand. However, some time ago, a survey showed that about 80% of software developed used about 20% of the instruction set of a CISC processor. This led to the development of the RISC processor.

RISC stand for **R**educed **I**nstruction **S**et **C**omputer. This was not a return to the technology of early processors. The space inside the IC which might otherwise have been occupied with circuitry to implement complex instructions could now be used to provide many more internal registers. Continually moving data between the processor and memory is a relatively slow process. If the data and any intermediate results can be left inside the processor whilst its being worked on, then considerable savings in time can be achieved.

In a further attempt to save time a process known as **pipelining** was also incorporated. This has some parallels with a production line in a factory. The first part of the pipeline fetches an instruction and passes it to the second part of the pipeline. If a set of instructions each required four steps and each step took one clock cycle then we would have:

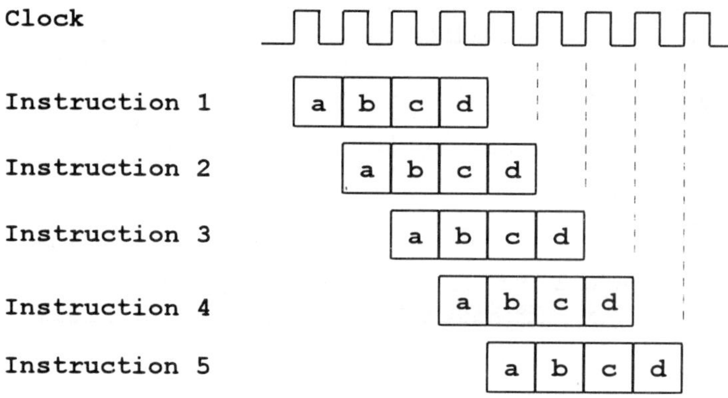

Figure 3.2 Pipelining

An instruction is commenced on each rising edge of the system clock and, although each instruction is shown as taking four clock pulses, the average time for an instruction is just one clock pulse. A further development of this are the **superscalar** processors. These have two or more pipelines operating in parallel and, on average, more than one instructions is executed every clock cycle.

You may remember from Chapter 1 that **operating speed** refers to the maximum speed at which the processor may be clocked. This has often been mistakenly interpreted as giving an indication of the speed at which the processor can execute a program or process data. If you compare superscalar processors, which may execute several instructions per clock cycle, with say a Z80, which takes ten clock cycles for an instruction such as LD A,(1234), you will see that operating speed is a very poor indication of the speed of the processor. It is only really justified when comparing processors of the same genus - a 4MHz Z80 for example can execute code at twice the speed of a 2MHz Z80.

3.4 SELECTING A PROCESSOR

At the beginning of any design process it is important to be sure about what it is you are trying to achieve. This should be given in the specification but invariably specifications are incomplete, particularly when they come from outside agencies who are looking for experts in a particular field. It is also important to realize that **the best design is the one which meets the specification at the lowest possible cost.** This does not mean "cheap and nasty". It is the specification which is important and if it is important that an item shall work reliably for twenty years then this forms part of the specification. A possible selection process is given below:

1. Make sure that you have the latest copy of the specification for the equipment.

2. Analyze the specification to see if there are any hidden or implied constraints.

3. Check on the quantity required and if the system is likely to be developed in the future.

4. Determine the requirements for the system in terms of memory, I/O, speed etc.. It may be helpful to classify the requirements in terms of vital, desirable and not important. Also identify any features which must bc avoided.

5. Select a number of processors which meet the requirements and make the final selection. At this stage you will be considering aspects such as cost, availability and reliability. (It is often worth favoring a processor which is already in use by the company since the cost of training staff can be quite high.)

3.5 PROCESSOR SELECTION FOR CASE STUDY 1

3.5.1 Initial specification

Refer to the first paragraph of case study 1 in Appendix A

3.5.2 Analyzing the specification

The phrase "general purpose training" implies that there is a free choice in the selection of a processor. This needs to be checked in case the request has come from an outside organization that really wants training on a specific processor.

We will assume that our original assumption was correct. We can now add some implied constraints to the specification e.g.

Education establishments are always complaining that there is not enough money for equipment and so the training board must be cheap.

Since it is required for "a good all round introduction to microprocessors" the students are likely to be novices and so the processor should be relatively simple.

The "good all round" part suggests that we will need to provide the basis for some more advanced work and so the processor can not be too simple.

The students will want to feel that they are using modern equipment and so some earlier processors which might seem ideal for teaching may have to be ruled out.

It is not clear in the original specification whether a stand alone system is required or whether something such as a PC may be used to download user programs. For the purpose of this exercise we will assume that a PC may be used.

3.5.3 Quantity and future development

There are a number of cheap processors available and so, whilst the quantity might affect the production process for the board as a whole, it will have little effect the choice of processor. Experience shows that systems of this kind are usually superseded by newer technologies long before upgrades might be introduced. However, to maximize the life of the board, we must try to select a processor which reflects both the current and the anticipated future trends in microprocessor technology.

3.5.4 System requirements

Determining the amount of memory required is difficult. We must either rely on experience to estimate the number of bytes required for each function, or we must look to other similar systems to make our estimate. Examining advertising literature shows that systems of this kind have operating systems which use between about 1K and 32K of memory. Those using the larger amounts of memory tend to include features which could reside on the PC. We are looking for a basic system so lets assume that we will be able to fit our operating system into about 4K of ROM.

The students will need RAM in which to enter their programs. Student exercises seldom use more than a few hundred bytes and so a 1K RAM should be more than adequate.

To demonstrate input and output will require a minimum of one output port and one input port.

Communicating with a PC is usually done through one of the computer's COMM ports. This is serial communication and so we must either use an extra serial I/O device or adapt the input and output ports. Again we are looking for cheapness so we will adapt an I/O port.

3.5.5 Processor Selection

For the purpose of this exercise we shall refer only to the summary of processor characteristics given in appendix E but the reader should remember that there are many more processors that might have been considered.

We can rule out several processors immediately : Cost eliminates the PowerPC 603 and later versions of the 80x86 and 68000 families of processors. The COP410L and PIC16C84 do not have enough memory. This leaves 8031, Z80, 8086 and 68000.

8031   **Advantages:** This processor and the rest of its family are very widely used for control applications. It has a modern architecture and a good selection of instructions for both elementary and more advanced work.
Disadvantages: The mixture of internal and external memory and the use of I/O ports for address and data bus may cause confusion with beginners.

Z80   **Advantages:** Low cost, good selection of instructions for both elementary and more advanced work.
Disadvantages: Although still widely used for control applications it is losing ground to other microcontroller families such as the 8031 and PICs. It may soon be perceived as out of date.

8086   **Advantages:** It is the simplest of the widely used 80x86 family and provides a good basis for work on the more advanced members of the family.
Disadvantages: The need to consider memory segmentation from the start may be confusing for beginners. Also it may be perceived as out of date.

68000   **Advantages:** It is the simplest of the 68K family, provides a good basis for work on more advanced members of the family, and has a good selection of instructions for both elementary and more advanced work.
Disadvantages: The bus structure may be confusing for beginners. Although widely used for control applications, its desktop computer role is being replaced by other processors and so it may be perceived as out of date.

At this stage it is tempting to rule out the Z80 as its long term future is not assured. For the final choice we need market research to establish if, in the long term, the majority of our customers will have a bias towards control or computing. If the bias is towards computing then the choice is between 68000 and 8086. In this case the 8086 is the clear winner due to the declining use of the 68000 in this area.

If the bias is towards control the choice is between 68000 and 8031. Technically there is little to chose between them for our application. The 8031 has probably got a marketing edge with people interested in control since it is a widely used processor, has a more modern architecture and includes on chip I/O which is characteristic of many micro controllers. It is therefore likely to have a better long term future for training purposes.

Exercise 3.1   - *** *suggested for portfolio* ***
Select and justify the choice of a processor for each of Case Studies 2 and 3.

4

Memory and I/O

You have already met the terms ROM, RAM and I/O. In this chapter we look at these in a little more detail. We will examine some of the different types and how they may be built into a system.

4.1 ROM

In earlier chapters the term ROM was used to indicate any memory whose contents could be regarded as more or less permanent and this is the way it is used by many authorities. More accurately however ROM refers to memory whose contents are fixed at the time the chip is manufactured. The cost of producing a ROM is determined partly by the cost to set up the manufacturing process and partly by the cost of running the process to make the chips. In general the setting up of the process is expensive but, once set up, the cost to manufacture each chip is low. For this reason true ROMs are usually only used where high volume production is anticipated. For many applications devices such as PROMs, EPROMs and EEPROMs described below offer a more economical solution.

4.2 PROM

PROM stands for **P**rogrammable **R**ead **O**nly **M**emory. PROMs are usually supplied to the user with all bits held at logical 0 by fusible links. If one of these links is blown (by passing a high current through it) the corresponding bit becomes a logical 1. The user will then use a piece of equipment known as a PROM programmer to blow selected fuses and thus enter his program into the device. Since the programming involves blowing fuses it is not possible to change the contents.

4.3 EPROM

EPROM stands for **E**rasable **P**rogrammable **R**ead **O**nly **M**emory. From a user's point of view this is similar to a PROM except that the program may be erased and a new program entered. This is particularly useful when developing new programs. The device has a quartz window and erasure is achieved by exposing the device to ultraviolet light.

The cost of putting in the window is quite high and this has led to another device with the unlikely and conflicting name of non-erasable EPROM. Internally it is identical to an ordinary EPROM but it has no window. Once the design team has solved both the electrical problems as well as the programming problems, a design may be transferred to the cheaper non-erasable EPROM without need for further changes.

4.4 EEPROM

EEPROM stands for **E**lectrically **E**rasable **P**rogrammable **R**ead **O**nly **M**emory. From a user's point of view this is similar to an EPROM but the contents of individual memory locations may be both erased and reprogrammed electrically. Also microprocessor based systems can be designed so that EEPROMs may be reprogrammed without the need to remove the chip from the circuit. The time taken to change the contents is rather long and so using EEPROMs as an alternative to RAMs is often not a practical nor, since they are much more expensive, an economic proposition.

EEPROMs have obvious advantages during system development. It should be remembered however that, if the ability to change the contents is left in the final design, there is the risk that other programs (say, user programs in RAM) may also alter the contents accidentally and thereby corrupt the operating system.

4.5 RAM: STATIC and DYNAMIC

Random access or read/write memory is designed for temporary storage of data and programs. Static RAM and dynamic RAM differ in the way they store information.

When a bit is written to a memory cell in static RAM a transistor is either switched on or off depending on whether the bit is to be a logical 1 or logical 0. Internal feedback is then holds the transistor in the required state. This means that the contents of static RAM will remain unchanged until either a new write operation is performed or the power to the circuit is switched off.

Dynamic RAM on the other hand holds information in the form of a charge on a capacitor. Capacitors are, by their very nature, leaky and, if allowed to do so, the charge will dissipate over a period of time. To avoid this, dynamic RAMs must be topped up or *refreshed*. In many systems this will mean adding extra circuitry for the purpose. The Z80 has a built in facility for refreshing dynamic RAM and this is put into action whilst the processor is decoding the op-code during the last part of each op-code fetch cycle.

The choice as to whether to use static or dynamic RAM will depend on the application. Dynamic RAMs are usually cheaper but extra cost is involved in providing refresh. Once they have been written to, they use less power than static RAMs. However, writing demands a high current and this may cause other problems. Static RAMs avoid the need for refresh but, since they always have at least one transistor conducting, there is a constant drain on the current supply.

4.6 VOLATILE AND NON-VOLATILE MEMORY

A memory device is said to be volatile if its contents are lost when the power is switched off and non-volatile if it retains its contents. ROM is clearly non-volatile and RAM is usually classed as volatile. There are, however, a group of devices referred to as non-volatile RAM. The most common of these are static RAMs with a built in battery which maintains the memory contents whilst power to the chip is removed.

4.7 I/O DEVICES

Input and output from a microprocessor based system is achieved with either standard logic ICs or with purpose built I/O chips. (Some microprocessor based systems use both.) Standard logic ICs such as tristate buffers and D'type latches (see Appendix F) tend to be used for simple input and output activities respectively where the direction of data flow is fixed and there is no additional formatting of the data required. Many of the purpose built devices may be programmed to be either input or output and may provide other features such as adjusting the rate at which data is sent or received and adding extra bits to ensure data integrity. (This will be dealt with in more detail later.)

4.8 MEMORY CHIPS: SIZE AND ORGANIZATION

Part of a catalogue description for a typical memory device might include statements such as:

> Size 16384 bits
>
> Organization 2048 words x 8 bits

This tells us that the device has 2048 memory locations and each memory can hold 8 bits. You might notice that 2048 x 8 = 16384 and this is the size or total number of bits that may be held in the device. To be able to identify 2048 different memory locations will require 11 address line since $2^{11} = 2048$ and, since each location holds 8 bits, there must be 8 data lines.

Now complete exercises 4.1 and 4.2 at the end of this chapter

4.9 ADDRESS DECODING

Address decoding refers to the way in which an address on the address bus is used to select the memory location to which it refers. In general, the higher order address lines are used to select a memory chip, and the low address lines are decoded inside the chip to select a particular memory cell. Figure 4.1 shows the layout of a microprocessor and its memory. We will use this to illustrate address decoding.

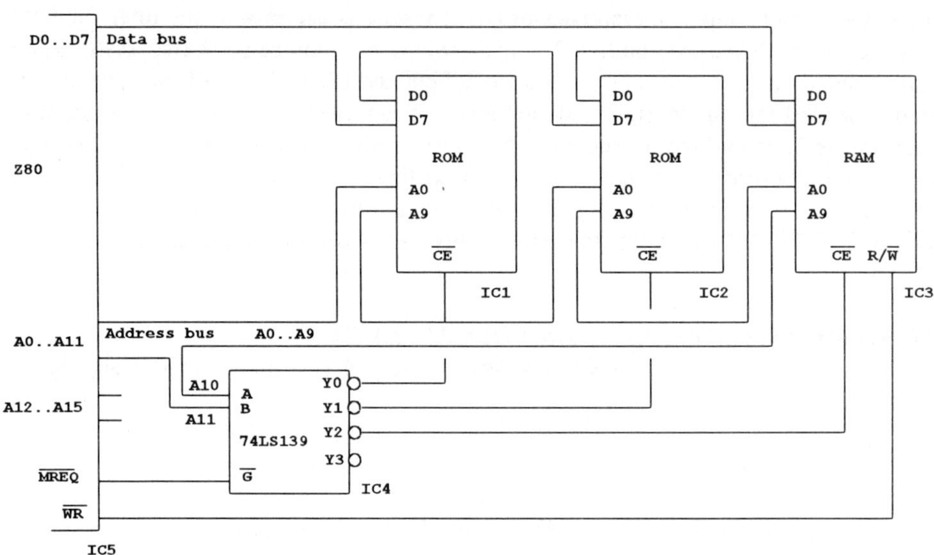

Figure 4.1

Operation:

1. Each memory chip may be regarded as having tristate buffers between its data pins and its internal circuitry. If the chip enable pin, **CE**, is high the buffers will be in the high impedance state and so the chip will be electrically isolated from the data bus even though it is still physically connected.

2. If CE is taken low the tristate buffers go to the active state and the chip is now electrically connected to the data bus (selected) and data can be exchanged between the processor and memory. It is clearly important that only one chip is selected at a time and this is the job of the 74LS139.

3. The operation of a 74LS 139 is shown by the table below

G	B	A		Y0	Y1	Y2	Y3
H	X	X		H	H	H	H
L	L	L		L	H	H	H
L	L	H		H	L	H	H
L	H	L		H	H	L	H
L	H	H		H	H	H	L

If the gate select pin, G, is high then all the output pins will be high what ever is on the inputs of A and B and hence non of the memory ICs will be selected. If the gate select pin is low one of the output pins will be low - which one will depend on the inputs to A and B.

4.   To see how the 74LS 139 selects a chip lets assume that the processor is doing a memory read from address 0420 hex. In binary the contents of the address bus will be:

```
0 0 0 0   0 1 0 0   0 0 1 0   0 0 0 0
              | |
    A11 __  |  |__ A10
```

a)   From the timing diagrams in chapter 2 we see that, for a memory read, ˙MREQ goes low. MREQ is connected to G and so one of the output pins on the 74LS139 will be low.

b)   Address line A11 is low and this is connected to B and so B will be low. Similarly A10 is high and so A will be high.

c)   The inputs to B and A are LH and, as we can see from the table above, this puts output Y1 low selecting IC2. The remaining memory chips have a logical 1 on their chip select pins and so remain in the high impedance state.

5.   Address lines A0 to A9 go to all three chips. Inside the memory IC these address lines are connected to circuitry which has a similar function to the 74LS139. This ensures that only one of the memory locations within the IC has access to the data bus.

If, for the moment, we assume that the address lines A12 through to A15 will always contain zeros, then we can determine the address range of each IC:

To select IC1 requires A11 and A10 to both be zero (i.e. **00**). Internally IC1 can decode any selection of 0's and 1's on A0 though to A9 and so the range of addresses for IC1 is

```
from     0000 0000 0000 0000        (0000 hex)
to       0000 0011 1111 1111        (03ff hex)
```

To select IC2 requires **01** on A11 and A10 and so its address range is:

```
from     0000 0100 0000 0000        (0400 hex)
to       0000 0111 1111 1111        (07ff hex)
```

Similarly for IC3 the address range is:

```
from      0000 1000 0000 0000        (0800 hex)
to        0000 1011 1111 1111        (0bff hex)
```

Manuals for microprocessor based systems often express address ranges in the form of a *memory map*. For the system in Figure 4.1 the memory map might appear as:

either   or

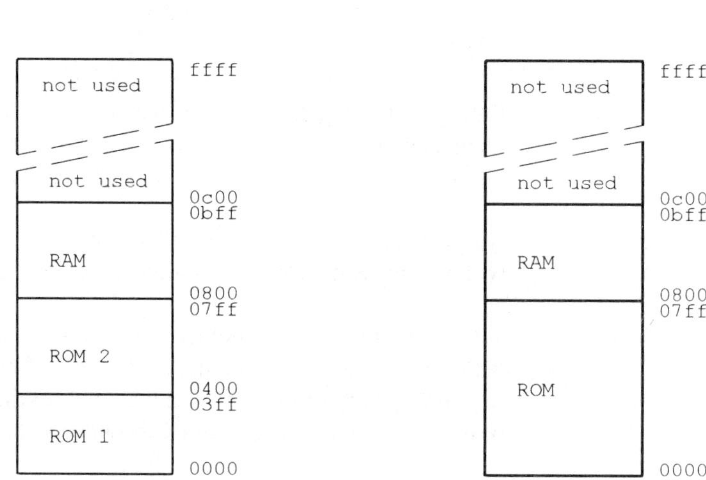

Earlier we assumed that, for this system, address lines A12 to A15 would be zeros. Let us now assume that, due to a programming error, the processor puts 8400 onto its address bus instead of 0400. In binary the address is 1000 0100 0000 0000. A10 is still a one, A11 is still a zero, A12 to A15 are not connected to anything and so IC2 will still be selected. Thus addressing 8400 instead of 0400 will still give access to address 0400. There are sixteen different combinations of zeros and ones which might appear on A12 to A15 and so there are sixteen different addresses at which the contents of memory location 0400 can be found. Indeed every memory location in this system has sixteen different addresses. This phenomenon is known as **memory ghosting.**

Memory ghosting will occur in any system which does not use all of its address lines in address decoding. If a system is designed which does not need all of the address lines then consideration must be given to the possibility that the system might be developed further in the future. If no further development is anticipated then memory ghosting may be acceptable. If it seems likely that future developments of the system will need more memory then steps should be taken to avoid memory ghosting. If you examine Figure 4.2 you will see that the unused address lines have now been taken to

an OR gate together with MREQ. Thus it is only when A12 to A15 and MREQ are all
low that a chip can be selected and so memory ghosting has been eliminated.

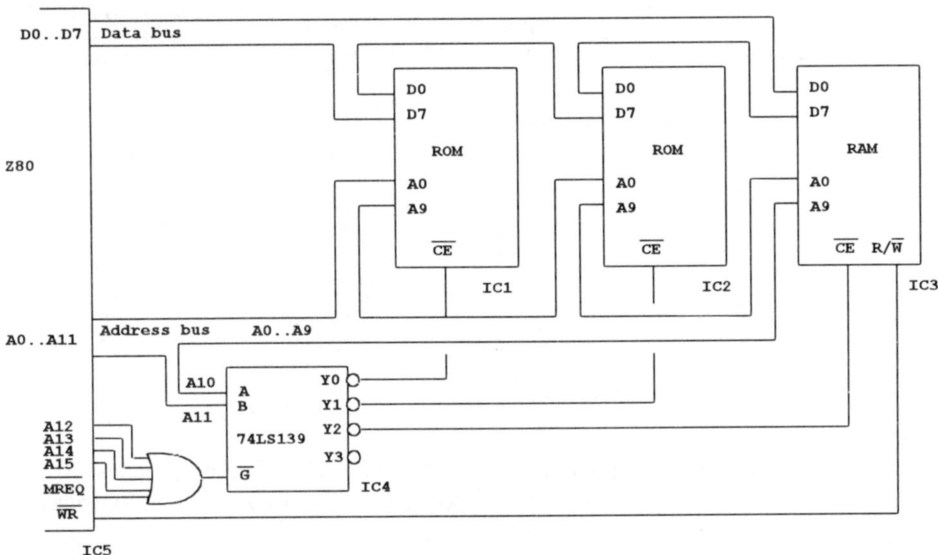

Figure 4.2

4.10 READ/WRITE CONTROL

With RAM chips and certain I/O chips it is necessary to control the direction of data
flow. Each data line on such devices may be thought of as having back to back tristate
buffers controlled by CE and R/W as shown in Figure 4.3.

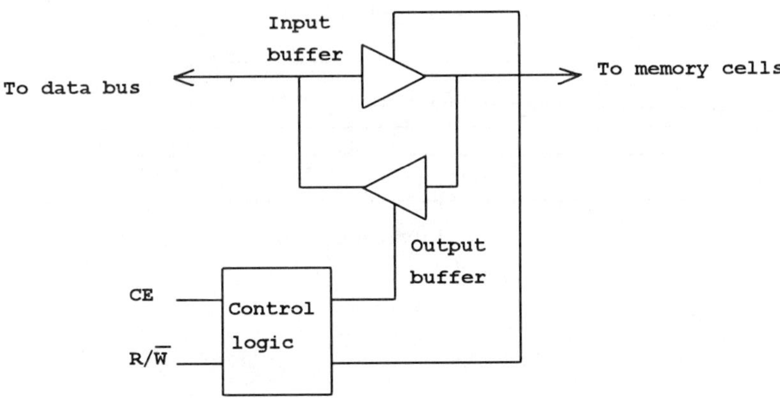

Figure 4.3

Read/write is controlled from the control logic as follows

CE	R/W	Input buffer	Output buffer	Overall operation
0	0	Active	High impedance	Memory write
0	1	High impedance	Active	Memory read
1	X	High impedance	High impedance	Chip not selected

4.11 ADDING I/O

No microprocessor system is any use unless it has some I/O. Figures 4.4 and 4.5 show
two different ways in which I/O might be added to the Z80 based system shown above.

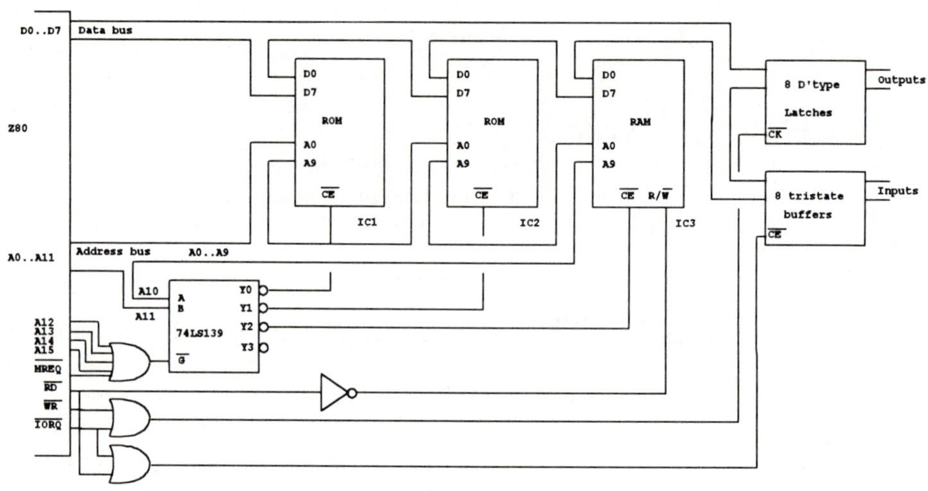

Figure 4.4

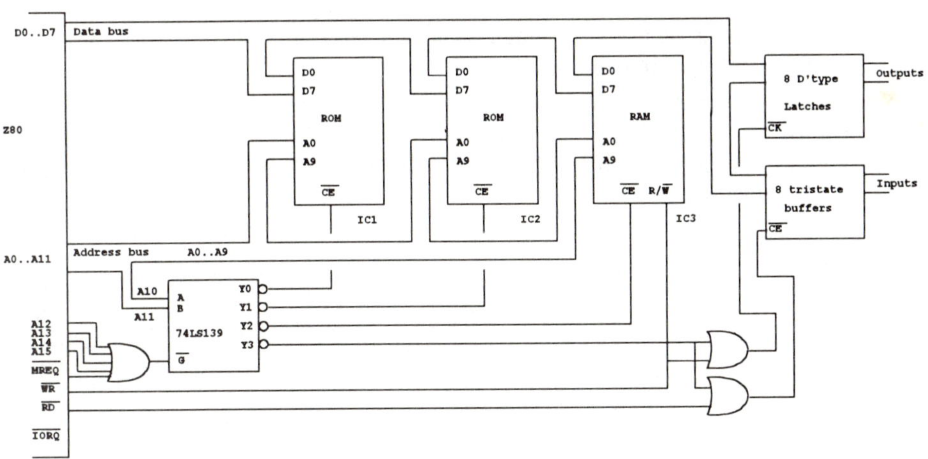

Figure 4.5

Notes 1.   Both systems use D'type latches for outputs and tristate buffers for inputs.

2.   In Figure 4.4 there are no address lines associated with the selection of either I/O device. Data on the data bus can be latched to the outputs by taking the CK pin low then high again. The CK pin is connected via an OR gate to IORQ and WR. If you examine the timing diagram for input and output instructions you will see that executing any OUT instruction will provide the necessary signal to latch data. In a similar way any IN instruction can be used to read data from the tristate buffers.

3.   Figure 4.5 illustrates **memory mapping** of an I/O device in a Z80 based system. The operation of either I/O device requires that output Y3 from the 74LS139 is low. This in turn means that A11 and A10 must both be logical 1. There are no other connections to the address bus and, since G on the 74LS139 is connected to MREQ, data can be latched to the ouputs by any memory write operation to binary address 0000 11XX XXXX. Similarly inputs can be read with any memory read operation from the binary address 0000 11XX XXXX .

Now complete Exercises 4.3 and 4.4 at the end of this chapter.

Exercise 4.1
The pin out diagram for a memory device shows address lines marked as A0 through to A11 inclusive and data lines marked I/O1 through to I/O8 inclusive.
     a)   Is this a RAM or a ROM?
     b)   What is the size and organization of the device?

Exercise 4.2 - *** *suggested for portfolio****
Examine some manufacturers catalogues and determine the range of size and
organization of readily available RAMs and ROMs

Exercise 4.3 - *** *suggested for portfolio****
Figure 4.6 shows a layout with a different processor. With reference to this diagram
determine the address range for each IC.

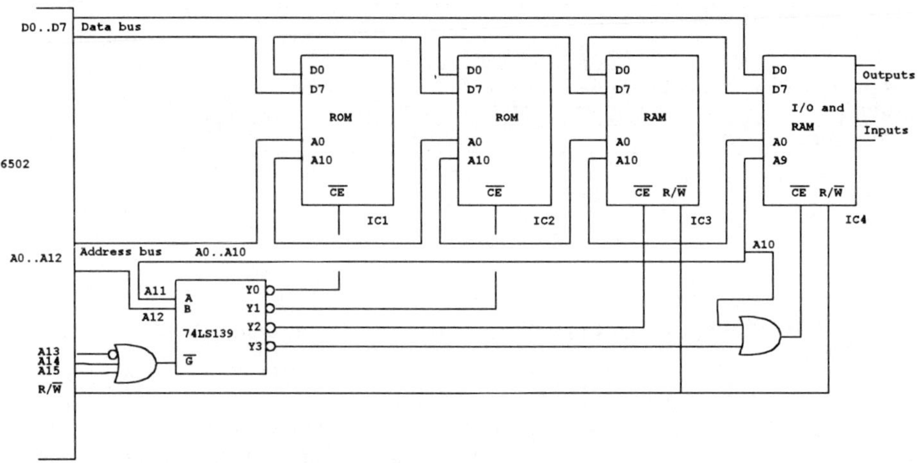

Figure 4.6

Exercise 4.4
The Z80 uses 8 bit addresses for I/O (A0 to A7). Show how the circuit of Figure 4.4
may be adapted so that the ouputs are only found at address 00 hex and the inputs are
only found at address 01 hex.

Exercise 4.5
Refer to Case Study 2. Write a short review which examines each type of memory
device mentioned in this chapter and assesses its suitability for use in the
microprocessor based control box.

5

Introduction to Programming Techniques

The best way to produce unreliable software which is difficult to debug, test and maintain is to start by writing code in the chosen final language.

In principle all problems associated with a program should be solved before any attempt is made to write code. Before a problem can be solved, however, we need to know that it exists and, once we think we have solved it, we need to be sure that our solution works. This has led to four main steps in program development:

Step 1 **Program specification.** It was mentioned in Chapter 3 that specifications, particularly those from outside organizations, are invariably incomplete. It is also true that, in an attempt to make them precise, the specifications become so long winded that interpretation is difficult and sometimes misleading. For Step 1 the program specification, as presented, should be analyzed and any hidden or implied constrains identified. It should then be set out as a series of short statements with each statement containing one fact about the specification.

Step 2 **Program design**. See notes below and Sections 5.1, 5.2 and 5.3.

Step 3 **Program implementation**. See Sections 5.4 and 5.5. This is the translation of the program design into the final chosen language. If Step 2 has been done properly, then, with a little experience, this should be a fairly simple task. However, if the program is very large, it may be necessary to develop it as several smaller programs which are combined into a single program at a later stage. This can cause additional problems particularly when writing in a low level language.

Step 4 **Program testing**. Looking back to the program specification gives a good guide as to the sort of tests that must be applied. In devising a test strategy it is important to ensure that, not only does the program do what its supposed to do, but also that it does not do something which it is not supposed to do.

Of these it is Step 2 - Program design which is the most abstract and is often omitted by the novice programmer. Unfortunately it is also the most important if we are to produce reliable software which is easy to maintain.

It is at the design stage that we decide how we are to meet the requirements of the specification. In the past many techniques have been used and, in general, they may be classed as either top down, bottom up, or ad-hoc. Each has merits for particular applications. By far the most widely used are, however, the top down and related techniques such as **JSP** (Jackson Structured Programming).

In general terms **top down design** (also known as **stepwise refinement, functional decomposition** and **hierarchical decomposition**) means splitting the initial problem into a series of smaller problems. Each of these sub-problems is then examined to see if a solution is possible. If a solution is not possible then the sub-problem is further divided and the process is continued until a complete solution becomes achievable.

As an example let us consider a school boy of my acquaintance who was a keen table tennis player. For a Design and Technology project he decided to build a machine which would continually serve balls to him so that he could practice his return of serve. A top down approach for this might begin as follows:

> *First try*: design a machine to continuously serve balls.

> *Second try*: we need a method of getting the balls to the bat and a method of making the bat serve the balls.

> *Third try*: for getting the balls to the bat we need something to hold a lot of balls and something to ensure that only one ball goes to the bat at a time; for serving the balls we need a method of aiming them, a method of projecting them and a method of ensuring that there is adequate time between serves.

> and so the process continues.

For a long time top down was seen as the way to develop programs. As programs grew larger, however, it became clear that still more structure was needed in the design process. For many people JSP fulfills this need.

Although JSP is claimed by its proponents to be a an entirely new approach to program design, it is in many ways just a top down approach which imposes a well defined structure on the program. JSP was originally developed for commercial data processing but it works quite well for engineering applications. There are three main steps to the JSP procedure:

> Step 1   Produce a data structure diagram which defines the structure of the data to be processed.

> Step 2   Translate the data structure diagram into a program structure diagram.

Step 3 Establish the operations and conditions for each part of the structure i.e.

a)   list the functional components of the program and assign them to the structure.

b)   list the conditions which influence selection and termination of iterations and add them to the program structure diagram.

To be able to use JSP it is essential that you are able to draw data structure diagrams and program structure diagrams. In the notes which follow simple examples are used which may at first appear to have little relevance to programming but they are useful in developing the concepts.

5.1 DATA STRUCTURE DIAGRAMS

A data structure diagram consist of elementary components shown as rectangular boxes containing the name of the component. The components may be grouped into control structures of which there are three types:

Sequence -   a series of two or more components which occur once each in a predetermined order.

Selection -   a series of two or more components of which, depending on the conditions, only one will be selected.

Iteration -   where a component may occur zero or a number of times.

Any horizontal grouping on the diagram may only have one type of control structure.

5.1.1 **Examples of Sequence** (note that the order left to right is the order in which they occur

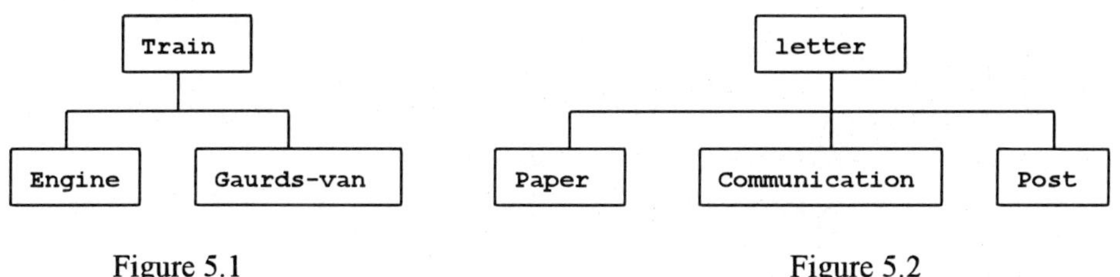

Figure 5.1   Figure 5.2

In Figure 5.1 train is composed of an engine and a guards-van. The engine appears first in the line because it is placed before the guards-van.

To write a letter (Figure 5.2) we need paper, we then need to write something on the paper and finally we need to post the letter.

If we are writing to a friend we might chose to include a photograph. In this case we need a new level as shown in Figure 5.3.

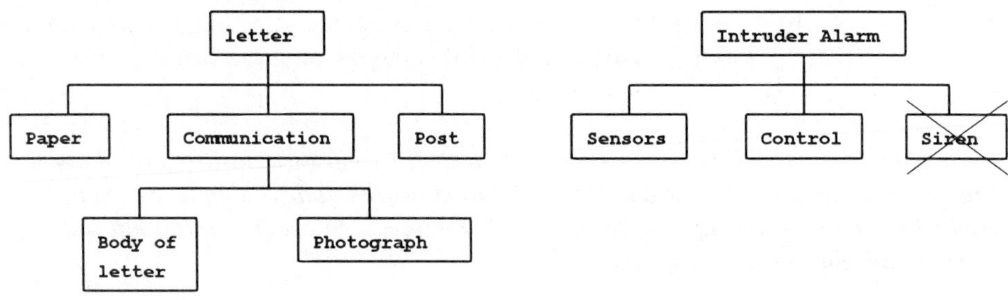

Figure 5.3   Figure 5.4

At first sight Figure 5.4 may appear satisfactory since data from the sensors goes to the control box and the control box determines whether the siren is to sound. However, depending on the output of the sensors, the siren may or may not be sounded and this is selection not sequence. Remember that any horizontal grouping may only contain one type of control structure and therefore this diagram is incorrect. You will see how to cope with this situation later.

5.1.2 Examples of Selection
Selection components consist of two or more sub-components of which only one is selected at any one time.   Since only one will be selected there is no significance in the order in which they occur. Selection components are shown in a similar way to sequence components but include a circle in the top right hand corner.

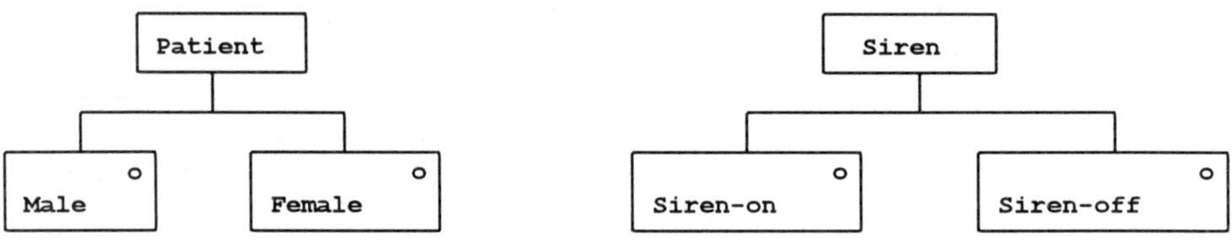

Figure 5.5   Figure 5.6

In Figure 5.5 a doctor's patient must be either male or female and in Figure 5.6 the siren must be either on or off. However there may be situations when either one, or the other or both are required. A company marketing domestic goods and services, for example, may find it useful to split its potential clients into male, female and couples. Since JSP allows only one component to be selected another box is required e.g.

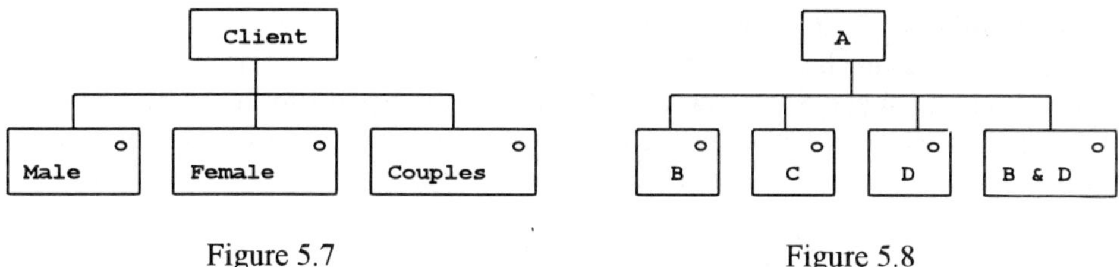

Figure 5.7   Figure 5.8

Figure 5.8 shows a more general situation in which either B or C or D or both B and D may be selected.

5.1.3 Examples of Iteration

Iteration refers to things which may occur zero or a number of times. Iteration components are shown with an asterisk in the top right hand corner.

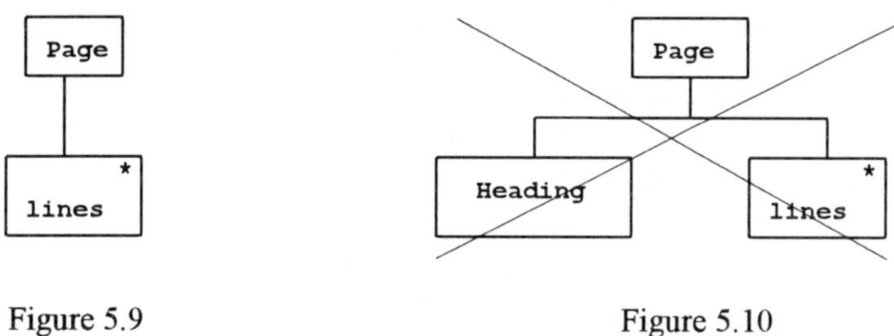

Figure 5.9   Figure 5.10

Figure 5.9 shows a page in a book which may be blank, it may have a few lines of text or it may have many lines of text. Sometimes it may be convenient to think of a page in a book as consisting of a header followed by lines of text. Figure 5.10 is an attempt to represent this on a data structure diagram but it is wrong because we are mixing two different control structures (sequence and iteration) in the same horizontal group.

Figure 5.11   Figure 5.12

Figure 5.11 shows the data structure for a college which runs a number of courses and each course has a number of students. In a similar way Figure 5.12 shows a doctors surgery from which a number of doctors operate and each doctor has a number of patients on his list.

5.1.4 Combining Control Structures

We have said that any horizontal group may consist of only one type of control structure and yet Figures 5.4 and 5.10 show an apparent need for mixed structures. The solution is an empty component (although you may give it a name for reference purposes) with the next structure at a lower level. Hence we could draw Figures 5.4 and 5.10 as follows

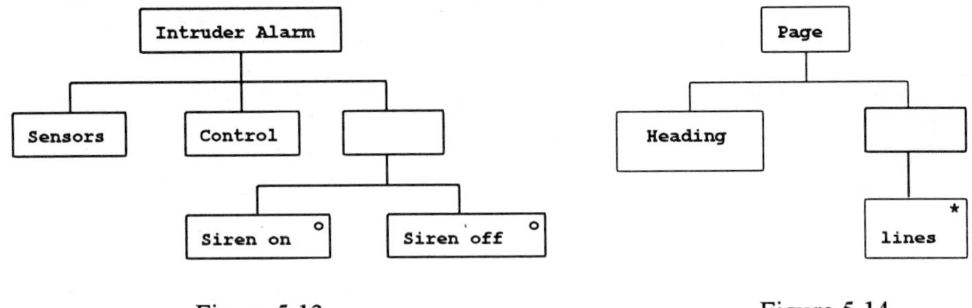

Figure 5.13   Figure 5.14

There is no absolute limit to the number of levels that may be included on a data structure diagram. Such limits that do appear are imposed by the nature of the problem. Student records held on a computer for example might have the following structure:

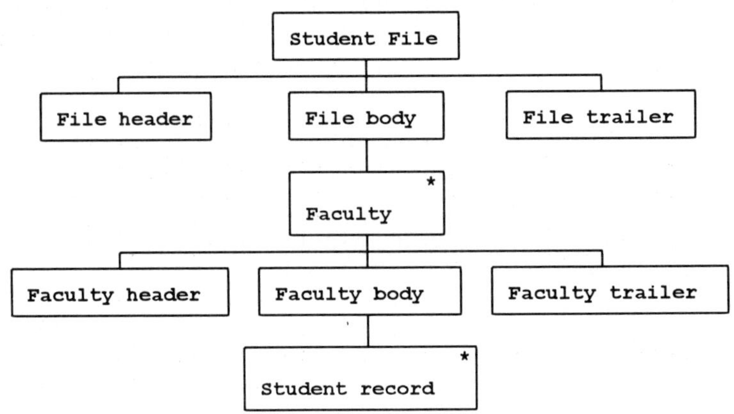

Figure 5.15

5.2 PROGRAM STRUCTURE CHARTS

For a program which processes only one set of data the construction of a program structure chart is a relatively straight forward process, since the program structure diagram will be either the same as the data structure diagram or a simple extension of it. When more than one data set is involved this will not be the case. Programs of this sort are, however, outside the scope of this book.

To indicate that a diagram relates to program structure and not data structure a letter P is added. Thus Figures 5.13 and 5.14 become:

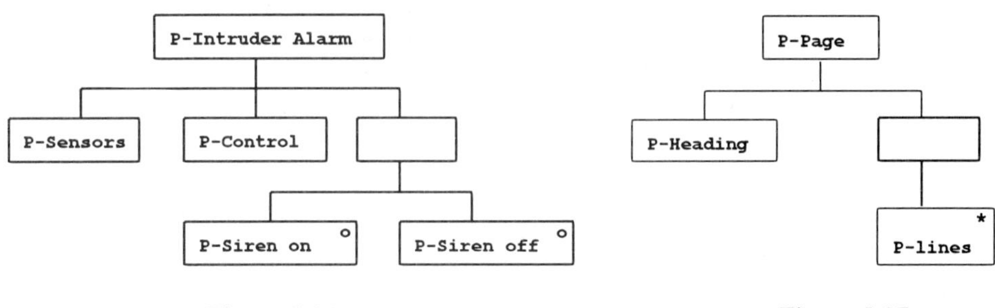

Figure 5.16   Figure 5.17

5.3 ALLOCATING OPERATIONS AND CONDITIONS

Once the program structure diagram has been drawn, it is examined and lists are drawn up for the operations required and conditions which influence transfer from one part of the diagram to another.

Example 5.3.1

A program is required to estimate the time required for a machinist to manufacture 25 items given the time to manufacture one item.

A possible data structure is shown in Figure 5.18 and a possible program structure in Figure 5.19

Figure 5.18   Figure 5.19

The operations needed are:

  1.   Enter the machinist name.
  2.   Enter the estimated time for each item.
  3.   Calculate the total time (= est. time x 25)
  4.   Output total time.

There are no conditions to allocate and so our final structure chart is:

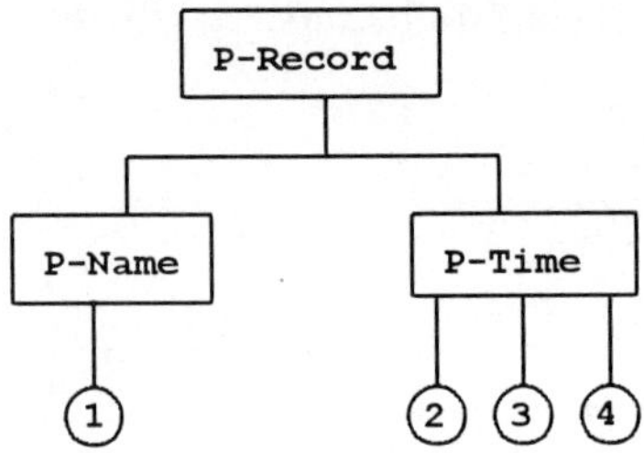

Figure 5.20

Example 5.3.2

If the machinist produces the items in less than three quarters of the estimated time a bonus is given. Amend Ex 5.3.1 to include an indication as to when the bonus is to be paid.

A possible data structure is shown in Figure 5.21

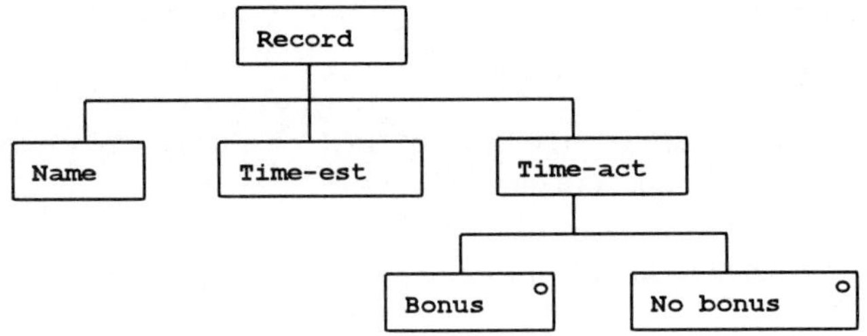

Figure 5.21

The operations needed are:

    1.   Enter the machinist name.
    2.   Enter the estimated time for each item.
    3.   Calculate the total time (= est. time x 25)
    4.   Calculate the time if a bonus is to be paid (= est. time x 25 x 0.75)
    5.   Output target time for machinist.
    6.   Input actual time taken.
    7.   Output message for bonus payment
    8.   Output message for no bonus payment

The conditions are:

    C1   Time fast enough for bonus
    C2   Time not fast enough for bonus.

Thus a program structure diagram could be:

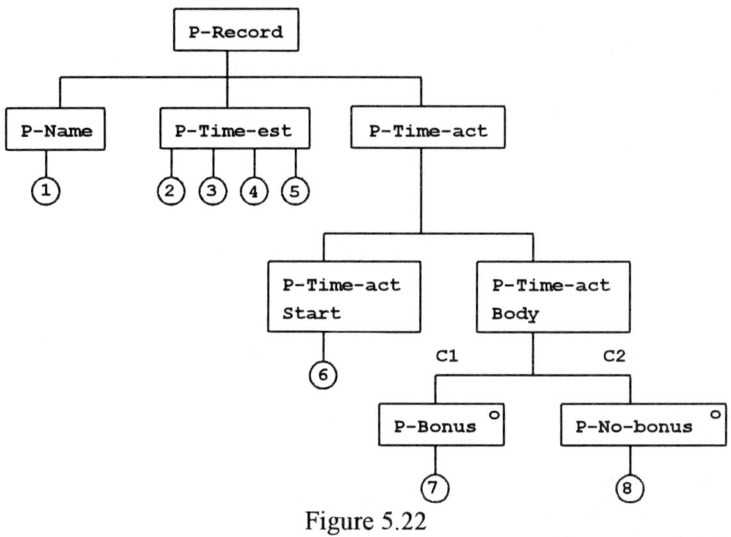

Figure 5.22

Now complete exercises 5.1 and 5.2 at the end of this chapter.

For program implementation all that remains is to translate the operations and conditions into code. For the beginner the easiest way of translating a program structure diagram into the final chosen language is to go via pseudo code. Indeed pseudo code is also the favorite route for many experienced programmers.

5.4 USE OF PSEUDO CODE

Writing pseudo code consists of first writing down the major steps required for a solution as a series of short statement that looks a bit like computer code (hence pseudo code). Each step is then examined to see if it can be translated into code: if not the step is subdivided. (Standards and conventions have evolved for pseudo code but we shall take a more relaxed approach.)

Example 5.2.1

An operation on a structure chart requires that two numbers are added together. It transpires that the numbers to be added are stored in memory locations 1234 hex and 1235 hex and the result is to be stored in memory location 5678 hex. Write three versions of the program: one each for Z80, 68000 and 8086.

Whatever processor is being used, addition is performed by the ALU. This means that we will probably need to get the numbers into the processor first. Thus our first attempt at writing pseudo code for the program might be:

    1.   Get the first number into the processor.
    2.   Get the second number into the processor.
    3.   Add the numbers.
    4.   Write the result back to memory.

We must now look to see if we have broken the problem down sufficiently for code to be written. When you have had just a little more experience you will know immediately that the answer is "yes". However, at this stage, you will need to look further. First we need to know if there is somewhere in the processor for both numbers. If you look at the diagrams for the internal structure of the processors you will see:

Z80   The accumulator A and any of the working registers B, C, D, E, H and L might be used.

68000   Any of the data registers D0 through to D7 might be used.

8086   Any of the registers AX, BX, CX and DX or the lower 8-bits of these AL, BL, CL and DL might be used.

Having decided that there is somewhere to put the numbers, we need to know if there are instructions which can be used to copy the data from memory to the registers. To establish this we must examine the instruction sets. The results of this examination are:

Z80   There is an instruction of the form **LD r,(nn)** which loads the contents of memory location **nn** into register **r**.

68000   There is an instruction of the form **move source,destination** which copies data from **source** to **destination**. Source may be a memory location and destination may be a register. The 68000 also requires us to specify the size of the data involved and, since we are moving a byte of data, the form becomes **move.b source,destination**

8086   There is an instruction of the form **mov reg,mem** which copies data from memory location **mem** to register **reg**.

Notice that, for each processor, both the abbreviations and the order in which the data appears is different. This is a result of manufacturers failing to agree on a common standard. We must however stick to these conventions since they have been adopted by manufacturers of software for converting instructions given in the form above (known as assembly code or assembler) into machine code.

Having decided that we can get the data into the processor we now need to know if we can add the numbers. Again if we examine the instruction sets we find:

Z80   There is an instruction of the form **add A,r** which adds the contents of register **r** to the accumulator (register **A**) and leaves the result in **A**. (Note that this means that one of the numbers must be in the accumulator.)

68000   There is an instruction of the form **add source,destination** which may be used to add the contents of source to the contents of **destination** and leave the result in **destination**.

8086   There is an instruction **add reg, reg** which may be used to add the contents of two registers and leave the result in the first named register.

The last operation is to write the instructions back to memory. Again examining the instruction set gives instructions of the form:

Z80   **ld (nn),r**
6800   **move.b source,destination**
8086   **mov mem,reg**

The program has therefore been subdivided sufficiently for code to be written. We must first decide which registers are to be used. Apart from the need to use the accumulator for one of the registers in the Z80 we are free to use any of those mentioned. Let us assume that we will use:

	First *number*	*Second* *number*	
Z80	**A**	**B**	) either upper case or
68000	**D0**	**D1**	) lower case may be used
8086	**AL**	**BL**	) for register names

When assembly code is written for either the Z80 or the 8086, hexadecimal numbers are indicated with a letter h (or H) after the number e.g. 1234h. When it is written for 68000, hexadecimal numbers are usually preceded with $ or occasionally h' (e.g. $1234 or h'1234)

Thus translating our pseudo code into assembler we have:

	Z80	**68000**	**8086**
Get first number	ld a,(1234h)	move.b $1234,d0	mov al,[1234h]
Get second number	ld b,(1235h)	move.b $1235,d1	mov bl,[1235h]
Add numbers	add a,b	add.b d1,d0	add al,bl
Write result to memory	ld (5678h)	move.b d0,$5678	move [5678h],al

To complete program implementation, all that remains is to use some appropriate software to convert the assembly code into a form which may be entered into memory. (Alternatively we could look up the codes in the instruction set but this is a rather long winded and error prone way of doing it.)

Notice that, so far, the pseudo code has been entirely independent of both the processor and the language being used to write the program. If you are writing your programs in high level languages such as BASIC, COBOL, C etc. this will always be true. When writing in assembly language, however, some thought needs to be given to the instructions available for the processor. If, for example, we had been asked to multiply the two numbers instead of add them the pseudo code might have been:

Get first number
Get second number
Multiply numbers
Write result back to memory

This would be satisfactory for the 68000 and the 8086 since both have instructions for multiplying numbers but no such instructions exist on the Z80. You may recall from your early school days that multiplication is a process of repeated addition i.e. 3 x 4 means add 3 to itself 4 times (3 + 3 + 3 + 3). To implement this we will need somewhere to store a running total and somewhere to hold a count of the number of times we have added. Thus to multiply two numbers on the Z80 pseudo code could be:

```
set count to zero
set total to zero

get first number
get second number
while count < second number
{
        add first number to total
        increment count by one
}
```

Some books would express the last part of this as:

```
loop while count < second number
        add first number to total
        increment count by one
end loop
```

We will deal with the implementation of this later. Notice though that this is still independent of the processor since, although it was written with the Z80 in mind, it could just as easily be used for either the 68000 or the 8086.

Note also the use of a blank line to separate initialization from the main task and the use of indents to make it easy to identify items within a loop.

Now complete Exercises 5.3 and 5.4 at the end of this chapter.

5.5 CONVERTING A STRUCTURE CHART INTO PSEUDO CODE

When converting a program structure chart into pseudo code, any sequence control structure may have the processes associated with it translated into consecutive lines of pseudo code, but any branch to a lower level will require a subprogram or subroutine. Consider for example Example 5.3.1 for which the program structure chart is given in Figure 5.20. The main program (P-Record) consists only of two sequence components (P-Name and P-Time) and their associated processes. Since this is entirely sequence the first attempt at pseudo code might be:

```
Main program P-Record
        get name
        get estimated time
        calculate total estimated time
        print result
        end
```

This is not yet good enough for writing machine code, but we have now imposed the program structure diagram onto our program.

Now let us look at Example 5.3.2 for which the program structure chart is given in Figure 5.22. The main program, P-Record, consist of three sequence components, P-Name, P-Time-est and P-Time-act. The first two consist only of processes 1 to 5 but the third involves a transition to a lower level and this means using a subroutine. Thus the first part of the translation process becomes:

```
Main program P-Record
        get name
        get estimated time
        calculate total estimated time
        print result
        call P-Time-act
        end
```

The subprogram P-Time-act consists of two sequence components P-Time-act Start and P-Time-act Body. P-Time-act Start has only one simple processes associated with it and so we can write the pseudo code directly but P-Time-act Body requires a further subprogram. The subprogram from P-Time-act requires a selection to be made to one of two further subprograms and so our final pseudo code might be:

```
Main program P-Record
        get name
        get estimated time
        calculate total estimated time
        calculate bonus time
        print result
        call P-Time-act
        end
```

```
Subprogram P-Time-act
        get actual time
        call P-Time-act-Body
        return


Subprogram P-Time -act-Body
        if  actual time < bonus time
                    call  Bonus
        else
                    call  No-Bonus
        return


Subprogram  Bonus
        print bonus message
        return


Subprogram No-Bonus
        print no bonus message
        return
```

The pseudo code above is a literal translation of the program structure diagram. Hopefully you are wondering why the program is so complicated when, by general inspection of the original problem, the pseudo code could have been:

```
Main program P-Record
        get name
        get estimated time
        calculate total estimated time
        calculate bonus time
        print result
        get actual time
        if actual time < bonus time
                print bonus message
        else
                print no bonus message
        end
```

5.6 TO JSP OR NOT TO JSP

The exercise above illustrates one of the main complaints of the opponents of JSP: it generates a lot of code. If you are writing programs that will sit on large machines where memory is not a problem, then this extra code is of little concern beyond its effect on the speed at which the data is processed. However, if the program is for a small dedicated control application, then the extra code may mean adding extra memory ICs and thereby increasing the cost. If the application is battery operated the extra memory is a drain on battery life. If the application involves the monitoring and control of high speed electrical signals, the reduction in speed may necessitate the use of hardware solutions instead of much cheaper software solutions.

Strict adherence to JSP principles imposes a well defined structure on the program. If the program is large and there is a complex data structure there can be little doubt that the assistance JSP gives to the development of reliable software which is easy to maintain, far outweighs the penalties of longer code. However, when the need arises to write tight code for control applications, it may be beneficial to start by generating data and program structure charts and then take a more liberal approach as to when subprograms are called.

Now complete Exercises 5.5 and 5.6 at the end of this chapter.

Exercise 5.1
Draw a data structure chart and a program structure chart for Case Study 1.

Exercise 5.2				- *** *suggested for portfolio* ***
Draw a data structure chart and a program structure chart for Case Study 2.

Exercise 5.3
Write the pseudo code for an operation to arrange the contents of memory locations 1234 hex and 1235 into ascending numerical order.

Exercise 5.4
Write the pseudo code for an operation which continuously monitors the temperature of a furnace. If the temperature is too low the heater is to be turned on, if the temperature is too high the heater is to be turned off.

Exercise 5.5
Write pseudo code for the interpretation of the program structure diagram for Case Study 1.

Exercise 5.6				- *** *suggested for portfolio* ***
Write pseudo code for the interpretation of the program structure diagram for Case Study 2.

6

Entering and Running Programs

You should by now have a reasonable grasp of the hardware associated with a simple microprocessor based system. All this hardware is quite useless without suitable programs telling the microprocessor what to do. Much of the remainder of this book is therefore concerned with the development of software for the system and the only way to develop an understanding of software is to write and test your own programs. To do this you will need to be familiar with the procedures for entering and running programs. This chapter looks at the basic procedure and how it needs to be adapted for a particular system.

Throughout we will be assuming that you will be basing your studies on a single processor which may be either an 80x86, a Z80 or a 68000. If it an 80x86 processor it will be assumed that you are doing all your work on a PC with an I/O board connected to the printer port. If it is either a Z80 or a 68000 processor it will be assumed that you will be entering your programs on a PC (sometime called the **host system**) and then sending it **(down loading it)** to the system which has the processor on which you are working **(the target system)**. The arrangement is shown in Figure 6.1.

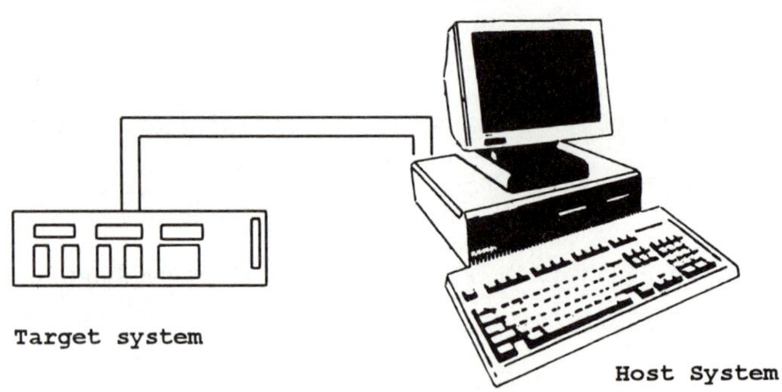

Figure 6.1

6.1 THE DEBUG CYCLE

Whatever system you are using the basic procedure for getting a program you have
written working is shown in Figure 6.2

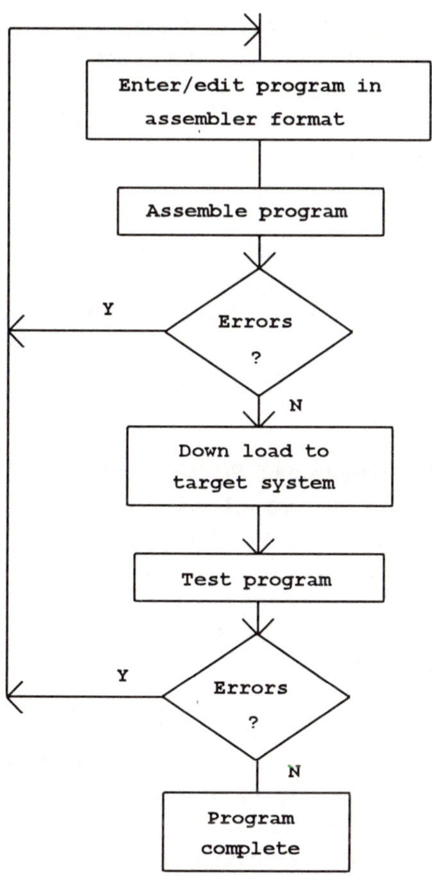

Figure 6.2

6.1.1 Entering programs in assembler format. Having designed your program using
JSP and pseudo code a word processor is used to enter the program in
assembly language format. This is then saved on disk as a text file and is often
referred to as a source file. Any word processor or text editor may be used to
create the source file provided that it can be made to produce pure ASCII code.
If you are not sure how to make your word processor produce pure ASCII code
you should consult the manual. Alternatively most PCs are supplied with a
text editor such as EDIT.EXE which may be used.

6.1.2 Assembling the program The microprocessor cannot understand programs entered in assembly language. The process of converting assembly language into a suitable format is known as assembling the program. In the early days of programming programs were **hand assembled**. This involved looking up the machine code equivalents of each assembly language instruction in the instruction set and was a time consuming and error prone operation. These days the majority of programs are **machine assembled**. For machine assembly a program called an assembler reads the text file containing your assembly language program and generates another file which may be sent to the processor

6.1.3 Errors (1) When you first assemble your program it is quite likely that it will have errors. The sort of errors that are detected at this stage are syntax errors. These are often simple typing errors such as hitting a comma when you wanted a full stop. Sometimes however invalid instructions will be reported. What ever the cause it will be necessary to go back and correct the source file then reassemble.

6.1.4 Down loading If you are using a PC with an I/O board attached to the printer port for all your work this step will not be necessary. However, depending on which assembler was used, it may be necessary to prepare the assembled file further before it can be used.

  If you are using a PC to assemble your programs and then down load them to a system with a different processor you should consult the manual for the system that you are using.

6.1.5 Testing It is important to remember that, although a program may have assembled correctly, it will not necessarily perform its intended task. The only thing about which you can be sure is that the program contains valid instructions. If you work on larger programs you should know that a lot has been written about test procedures. For the purposes of this course, however, testing will be limited to checking that the program performs its intended task on a limited selection of data.

6.1.6 Errors (2) If testing detects errors in the program then it is necessary to go back and amend the source file. The file must then be reassembled and any new syntax errors removed before testing the program again.

N.B.   The examples which follow are intended to give you practice in entering, running and altering programs. It is not intended that you should work through all the examples. The choice as to which you use will be determined by the hardware and software that you are using.

For those using one of the 80x86 family of processors
In examples 6.1, 6.2, 6.3 and 6.10 it is assumed that you will be running the program on a PC and will be using Microsoft Macro Assembler 5.1.

Example 6.1
This program should cause a tone to be generated on the speaker in the PC.

Step 1   Invoke the text editor with:

```
edit   tone.asm              <cr>
```

edit is the name of the editor.
tone is the name of the source file. Any combination of up to eight letters and/or numbers may be used.
.asm is the file type or file extension. For most assemblers it is conventional to use .asm and some will accept nothing else.
<cr> indicates that you should hit the carriage return key

Type in Program 1 shown below paying particular attention to the following points:

1.   Words such as speaker, freq and start: which appear in the first column of some lines must also occur in the first column in your program.
2.   Words such as *TITLE, .DATA* and *mov* should start two tab positions from the left hand edge.
3.   If you are using a 386 machine or later the program should give a satisfactory results as it is. If you are using an earlier machine you will get better results if you replace

```
freq      dw      60000
dura              dw      10
step_sz  dw      200
```

with

```
freq      dw      10000
dura              dw      20
step_sz  dw      100
```

```
;******************************************
; Program 1 - practice in entering and       *
; running programs on an 80x86 running DOS *
;******************************************
                TITLE     tone
                DOSSEG
                .MODEL    SMALL
                .STACK    100h                    ; allocates 256 bytes for stack

                .DATA                             ; marks start of data segment

freq            dw        60000
dura            dw        10
step_sz         dw        200

speaker         equ       61h

                .CODE                             ; marks start of code segment

start:          mov       ax,@data                ; loads segment location
                mov       ds,ax                   ; into DS register
                cli

                in        al,speaker
                and       al,0feh
                mov       cx,freq

next_freq:      mov       dx,dura
again:          xor        al,2
                out       speaker,al
                mov       cx,freq
wait1:          loop      wait1
                dec       dx
                jnz       again
                mov       cx,freq
                sub       cx,step_sz
                mov       freq,cx
                jne       next_freq
                sti

                mov       ax,4c00h                ; Used to terminate a program
                int       21h                     ; and return to DOS

                END       start
```

Step 2   Assemble your program with the command

```
masm tone,,,,;                <cr>
```

The four comas and a semi colon may seem a bit odd but it forces masm to produce certain intermediate files that are needed for the next stage. (If your file has a different name then you replace ***tone*** with the name of the file.) If you issue the command

```
dir tone.*                <cr>
```

you will see that, in addition to *tone.asm* there are now another three files called *tone.lst, tone.obj* and *tone.crf*.)

If there are any errors reported you must correct the source file and reassemble. (If you manage to complete the whole exercise with no errors you might find it benificial to include some deliberate errors to see how they are reported.)

Step 3   The program must now be prepared for use with DOS. This means using a linker. Larger programs are usually split into several files and the job of the linker is to link these files together to form a single program. Since we are only using a single file the use of a linker seems a bit of an overkill but it is the simplest way to make this software produce a program which will run with DOS. Issue the command:

```
link tone,;                <cr>
```

(Again you can replace ***tone*** with the file name of your program.)

Step 4   Run the program by issuing the command

```
tone                      <cr>
```

Example 6.2
Now change your 8086 program to that shown below. The differences are shown in bold italic type. Reassemble, link and run the new program.

```
TITLE        hello
DOSSEG
.MODEL   SMALL
.STACK   100h

.DATA
```

```
start_freq    dw        60000           ; *see note below
freq          dw        0
dura          dw        2
step_sz       dw        2000
num_cyc       dw        10


speaker       equ       61h


              .CODE


start:        mov       ax,@data
              mov       ds,ax
              cli

              in        al,speaker
              and       al,0feh

cyc_start:    mov       cx,start_freq
              mov       freq,cx


next_freq:    mov       dx,dura
again:        xor       al,2
              out       speaker,al
              mov       cx,freq
wait1:        loop      wait1
              dec       dx
              jnz       again
              mov       cx,freq
              sub       cx,step_sz
              mov       freq,cx
              jne       next_freq
              dec       num_cyc
              jne       cyc_start
              sti

              mov       ax,4c00h
              int       21h

              END       start
```

* Again if you are using an older machine the following values might be more suitable

```
start_freq    dw        10000
freq          dw        0
dura          dw        10
step_sz       dw        500
```

Example 6.3
Experiment with other values of dura and step_sz to create different effects.

Example 6.4

The following program should cause three of the LEDs on the I/O board to illuminate
one after the other. Experiment with this program to establish the effect of changing
del_time.

```
                TITLE          hello
                DOSSEG

                .MODEL  SMALL
                .STACK  100h          ;allocates 256 bytes for
stack

                .DATA                 ;marks start of data segment

o_port          dw      0
i_port          dw      0
del_time        db      10h

                .CODE                 ;marks start of code segment

start:          mov     ax,0
                mov     ds,ax
                mov     bx,ds:[408h]

                mov     ax,@data   ; loads segment location
                mov     ds,ax      ; into DS register
                mov     o_port,bx
                inc     bx
                mov     i_port,bx
                mov     dx,o_port
next:           mov     al,80h
                out     dx,al
                call    delay
                mov     al,40h
                out     dx,al
                call    delay
                mov     al,20h
                out     dx,al
                call    delay
                jmp     next
```

```
delay:          mov     al,10h
loop3:          mov     bl,0h
loop2:          mov     cl,0h
loop1:          dec     cl
                jne     loop1
                dec     bl
                jne     loop2
                dec     al
                jne     loop3
                ret
                mov     ax,4c00h   ; Used to terminate a program
                int     21h        ; and return to DOS

                END     start
```

For those using a Z80 processor

Examples 6.5, 6.6 and 6.7 assume that you are using a Z80 based target system which has some free user RAM starting at address 1800 hex and a Z80 PIO with LEDs attached to port B at I/O address 81 hex. You should check the manual to ensure that these details are compatible with the system you are using. If your system has user RAM at other addresses you will need to change

```
        .org    1800h
```

to reflect the new address. If the address of the LEDs is different you will need to change

```
led     .equ    81h
```

*to reflect the new address. The examples also assume that you will be using the Shareware Table Assembler TASM.EXE. If you are using a different assembler you will need to check the manual - the most likely differences will be the omission of the full stops to make **org** instead of **.org** and **equ** instead of **.equ**.*

Example 6.5

The following program should cause three of the LEDs on the I/O board to illuminate one after the other.

Step 1   Invoke the text editor with:
```
        edit    leds.asm                <cr>
```

> **edit** is the name of the editor.
> **leds** is the name of the source file. Any combination of up to eight letters and/or numbers may be used.
> **.asm** is the file type or file extension. For most assemblers it is conventional to use .asm and some will accept nothing else.
> **<cr>** indicates that you should hit the carriage return key

Type in Program 2 shown below paying particular attention to the following
points:

1. Words such as led and start: which appear in the first column of a line
    must also occur in the first column in your program.
2. Words such as *.org, .equ* and *ld* should start two tab positions from
    the left hand edge.
3. The remaining data such as 1800h and a,4fh should start one further
    tab over.

```
;**********************************
; Program 2 - practice in entering *
; and running programs on a Z80     *
;**********************************
;

             .org      1800h

led          .equ      81h
led_ctrl     .equ      led+2

del_time     .equ      40h

        ; ***** set up pio *****

             ld        a,4fh
             out       (sw_ctrl),a
             ld        a,0fh
             out       (led_ctrl),a

        ; ***** main program *****

start:       ld        a,80h
             out       (led),a
             call      delay
             ld        a,40h
             out       (led),a
             call      delay
             ld        a,20h
             out       (led),a
             call      delay
             rst       30h
```

```
delay:          ld          b,del_time
loop2:          ld          a,0
loop1:          dec         a
                jp          nz,loop1
                dec         b
                jp          nz,loop2
                ret


                .end        1800h
```

Step 2   Assemble your program with the command

```
tasm -80 -s leds.asm leds.hex leds.lst leds.sym <cr>
```

(If your file has a different name then you replace *leds* with the name of the file.)

The command line for using tasm is rather long winded. If you are going to be using TASM often you can reduce the amount of typing by creating a batch file

e.g. Invoke editor with:
```
        edit asm80.bat
```

Enter:
```
tasm -80 -s %1.asm %1.hex %1.lst %1.sym      <cr>
```

Then save the file. It is now possible to invoke the assembler with the full command line by typing:
```
        asm80 filename
```
e.g.
```
        asm80 leds
```
         (note extension *.asm* is ommitted)

If there are any errors reported you must correct the source file and reassemble.

Example 6.6
Now alter your Z80 program by changing:

        rst   ***30h***

to

        jp   ***start***

Then reassemble your program and run it again.

Example 6.7
Experiment with different values for del_time. Try to create the effect of a running light. If you make the value of del_time small enough, but not zero, it should appear as though all three lights are permanently illuminated. Try making del_time equate to zero - this is very small but the effect may not be quite what you expect.

For those using one of the 68000 family of processors

Examples 6.8, 6.9 and 6.10 assume that you are using a 68000 based target system which has some free user RAM starting at address 400400 hex and a 68230 Peripheral Interface/Timer whose base address is at 800001 hex and has LEDs attached to port B. You should check the manual to ensure that these details are compatible with the system you are using. If your system has user RAM at other addresses you will need to change

```
        org   $400400
```
to reflect the new address. If the base address of the 68230 is different you will need to change the line

```
        pit   equ $800001
```
to reflect the new address. If the LEDs are connected to port A then change:

```
        pit+$12      to    pit+$10
        pit+$6       to    pit+$4
        pit+$e       to    pit+$c
```

The examples also assume that you will be using an assembler similar to the Flight Electronics' produce XASM.EXE. If you are using a different assembler you will need to check the manual - the most likely differences will be the inclusion of full stops with the pseudo-ops to make **.org** instead of **org** and **.equ** instead of **equ**. The other difference sometimes encountered is that hexadecimal values are indicated with h' instead of $ e.g.

```
        org   h'400400
```

instead of
```
              org   $400400
```

Example 6.8
The following program should cause three of the LEDs on the I/O board to illuminate one after the other.

Step 1   Invoke the text editor with:

```
        edit   leds.asm               <cr>
```

edit is the name of the editor.

leds is the name of the source file. Any combination of up to eight letters and/or numbers may be used.

.asm is the file type or file extension. For most assemblers it is conventional to use .asm and some will accept nothing else.

<cr> indicates that you should hit the carriage return key

Type in the Program 1 shown below paying particular attention to the following points:

1.	Words such as led and repeat which appear in the first column of a line must also occur in the first column in your program.
2.	Words such as org, equ and move.b should start two tab positions from the left hand edge.
3.	The remaining data such as **#$0ff,led_ddr** should start one further tab over.

```
************************************
* Program 3 - practice in entering  *
* and running programs on the 68000 *
************************************


pit             equ        $800001
led             equ        pit+$12
led_ddr         equ        pit+$6
led_cr          equ        pit+$e
del_time        equ        $40000

                org        $400400

start           clr.b      led_cr
                move.b     #$0ff,led_ddr

repeat          move.b     #$80,led
                bsr.s      delay
                move.b     #$40,led
                bsr.s      delay
                move.b     #$20,led
                bsr.s      delay
                trap       #11
                dc.b       0,0
```

```
delay              move.l    #del_time,d0
loop1              sub.l     #1,d0
                   bne.s     loop1
                   rts


                   end       start
```

Step 2 Assemble your program with the command:

```
xasm leds.asm
```

Step 3 Refer to the manual for your system and download the assembled program to the target system.

Step 4 Refer to the manual for your system and run the program.

Example 6.9
Now alter your 68000 program by replacing the lines

```
          trap      #11
          dc.b      0,0
```

with a single line
```
          bra.s     repeat
```

Then reassemble your program and run it again.

Example 6.10
Experiment with different values for del_time. Try to create the effect of a running light. If you make the value of del_time small enough, but not zero, it should appear as though all three lights are permanently illuminated. Try making del_time equate to zero - this is very small but the effect may not be quite what you expect.

6.2 USE OF LABELS AND CONSTANTS
A *label* is a name given to a line of code which after assembly will reflect an address in memory e.g. in example 6.4 we have a line:

```
delay:          ld          b,del_time
```

'delay:' is a label and if you compare it to the program below you can see that it corresponds to address 181C hex. Once an address has been labeled it can be referred to by its label e.g.
```
          call      delay
```
A *constant* is a group of characters which are used to represent a value to be used in the program e.g. in example 6.4 we have a line

```
del_time       .equ       40h
```

'del_time' is a constant. When a constant has been equated (hence .equ) to something we can refer to it by name instead of value.

We can illustrate the benefits of using labels and constants by comparing the program in example 6.4 with the same program when it is written without labels and constants as shown below:

```
                      ; set up pio

1800 3E 0F                    ld         a,0fh
1802 D3 83                    out        (83h),a


                      ; main program

1804 3E 80            ld         a,80h          ;labeled start in ex
6.4
1806 D3 81            out        (81h),a
1808 CD 1C 18         call       181ch
180B 3E 40            ld         a,40h
180D D3 81            out        (81h),a
180F CD 1C 18         call       181ch
1812 3E 20            ld         a,20h
1814 D3 81            out        (81h),a
1816 CD 1C 18         call       181ch
1819 C3 04 18         jp         1804


181C 06 40            ld         b,40h          ;labeled delay in ex
6.4
181E 3E 00            ld         a,0
1820 3D               dec        a              ;labeled loop in ex 6.4
1821 20 FD            jr         nz,1820h
1823 05               dec        b
1824 20 FA            jr         nz,1820h
1826 C9               ret
```

Comparisons

1.   Using labels and constants makes the program much easier to follow.
2.   The constant *led* appears three times and in longer programs it may appear many more times. If it becomes necessary to convert this program for use on a system which has a PIO at a different address then, using constants only one line needs to be changed:

```
led                   .equ       81h
```

The corresponding change to **led_ctrl** occurs automatically because of the line

```
led_ctrl      .equ      led+2
```

Without the use of constants it would be necessary to change every line on which the constant occurred

3.   Similar advantages apply to labels. In addition if, during program development, lines of code need to be added or removed addresses will change. This will also require every reference to these addresses to be changed. In large programs the chances of missing one or more changes are quite high. Using labels, however, we can leave the task to the assembler and be sure that nothing has been missed.

6.3 PSEUDO-OPS

Pseudo-ops are instructions for the assembler rather than the processor. The pseudo-ops are not usually case sensitive (i.e. EQU is treated as the same as equ) but the range available and the precise way in which they are expressed may vary from one assembler to another. In this book it will be assumed that you are using one of the assemblers mentioned in section 6.1 and we will restrict our selves to the following:

EQU (8086 and 68000) or **.EQU** (Z80)
This assigns a value or set of characters to a constant as in **led .equ 81h**

ORG (8086 and 68000) or **.ORG** (Z80)
For the 68000 and the Z80 this tells the assembler the address of the first byte of code. The assembler then uses this to calculate all of the other addresses. Its use with MASM is rather more complex but will not be needed for the examples in this book.

END (8086 and 68000) or **.END** (Z80)
This indicates to the assembler that there is no more data or code to be included. With the 68000 and the Z80, END is usually followed by a reference to the label of the first byte of code e.g. **END start**. This gives additional information to the assembler as to how it should prepare the file which is to be down loaded to the target system.

DB and **DW** (8086) or **.BYTE** (Z80) or **DC.B**, **DC.W** and **DC.L** (68000)
These are usually used in conjunction with a label and are used to indicate to the assembler that the rest of the information on the line is data rather than code. DB, .BYTE and DC.B refer to byte size data, DW and DC.W refer to word size data (2 bytes) and DC.L refers to long word size data (4 bytes).

OFFSET (8086)

This returns the address of a label. If the data segment of an 8086 program contained the line

```
num1            db       56h
```

Then the line

```
        mov     bx,OFFSET num1
```

would put the **address** of num1 into register bx **not** the value 56h.

6.4 SKELETON FILES

If you are going to be using one particular system with the same I/O facilities for all your programming examples it is worth while creating a skeleton file which contains all the code that is likely to be needed every time. If, for example, I was going to do a lot of work on a Z80 based system which had a Z80 PIO with switches attached to port A at address 80h and LEDs attached to port B at address 81h I would prepare a file called **Z80_skl.asm**. This file would contain the following:

```
;********************************
; skeleton file for Z80 programs *
;********************************
;

              .org   1800h

port_a        .equ   80h
sw            .equ   port_a
led           .equ   port_a+1
sw_ctrl       .equ   port_a+2
led_ctrl      .equ   port_a+3

;**** add user equates between these lines ****

;**************************************************
;

start:        ld     a,4fh
              out    (sw_ctrl),a
              ld     a,0fh
              out    (led_ctrl),a

;**** add user code between these lines *******

;**********************************************
;
exit:         rst    30h
              .end   start
```

Then, if we wanted to write a program to say show a binary count, we would use the DOS copy command to copy the skeleton file and give it a relevant name e.g.

```
copy z80_skl.asm bin_cnt.asm
```

we could then call up the editor with the command:

```
edit bin_cnt.asm
```

and alter the file to meet our needs. The way we have expressed the equates may seem a little odd but it has been done for the following reasons:

1.   All PIO addresses are referenced back to port A. If we need to move the program to a system with a PIO at different addresses there is only one line to change i.e.

```
        port_a          equ     80h
```

2.   By including the lines:

```
        sw              equ     port_a
        led             equ     port_a+1
        sw_ctrl         equ     port_a+2
        led_ctrl        equ     port_a+3
```

we can forget about whether the switches are connected to port A or port B and refer to them by a convenient name.

3.   Although not strictly part of this course you should be aware that ports on a Z80 PIO may be programmed to be either inputs or outputs. Each port has a control register which determines which bits on a port are to be inputs and which are to be outputs. To make all bits on a port inputs it is necessary to send 4fh to the port control register. Similarly, to make all bits outputs, it is necessary to send 0fh to the port control register. It is much quicker to delete unwanted lines than to type wanted lines in and so the following lines were included:

```
start:          ld      a,4fh
                out     (sw_ctrl),a
                ld      a,0fh
                out     (led_ctrl),a
```

If you are using a 68000 system with a 68230 at base address of $800001 I would suggest the following skeleton file:

```
*********************************
* Sleleton file for 68000 programs *
*********************************


        pit         equ         $800001
        sw          equ         pit+$10
        sw_ddr      equ         pit+$4
        sw_cr       equ         pit+$0c
        led         equ         pit+$12
        led_ddr     equ         pit+$6
        led_cr      equ         pit+$0e


***** add user equates between these lines ****


*************************************************


                    org         $400400

start               move.b      #$80,sw_cr
                    move.b      #$0,sw_ddr
                    move.b      #$80,led_cr
                    move.b      #$0ff,led_cr


***** add user code between these lines *******


*************************************************
exit                trap        #11     * check your manual for
                    dc.b        0,0     * best way to terminate
                                        * program on your system


                    end         start
```

If you are going to be doing your programming exercises on a PC and use the printer
port for I/O you will be faced with two problems. The first is that not all PCs have the
printer port at the same address and second is that, because of the way DOS operates,
you can not guarantee to have your programs in any particular area of memory.

The first problem is easy to overcome. When you first turn on the computer it
checks to see where the printer ports are located and stores the first address (outputs) of
the first printer port (LPT1) at memory location 0:408h. Thus we can include a section
in our program which reads this address, determines the address for inputs and stores
both addresses in the code section of our program for future use.

To solve the second problem we must use only labels for addresses and let
masm.exe, link.exe and DOS sort out the addresses for us. If we need to know the

addresses for debugging purposes we must use single step and look at register contents - this will be dealt with later.

These problems leave us with a rather lengthy skeleton file but, once written, it is straight forward to use.

```
************************************************************
* Sleleton file for 8086 programs to be used with masm *
************************************************************

            TITLE         skeleton
            DOSSEG

            .MODEL   SMALL
            .STACK   100h        ;allocates 256 bytes for stack

            .DATA                 ;marks start of data segment

o_port      dw       0
i_port      dw       0

;****** add user data between these lines *****

;*******************************************

            .CODE                   ;marks start of code seg

start:      mov      ax,0
            mov      ds,ax          ;point ds at segment 0
            mov      bx,ds:[408h]   ;read port address

            mov      ax,@data       ;loads user data segment
            mov      ds,ax          ;location into DS reg

            mov      o_port,bx      ;store O/P port address
            inc      bx             ;calc I/P port address
            mov      i_port, bx     ;store I/P port address

;****** add user code between these lines *****

;*******************************************

exit:       mov      ax,4c00h       ;terminate program
            int      21h            ;and return to DOS

            END      start
```

7

Assembly Language Programming - 1

In learning to write programs in assembly language it is important that you write and test your own programs. To do this you will need to concentrate on one processor but, for future development, it is also important that you recognize similarities with other processors. This book deals with three processors 8086, Z80 and 68000 and it is recommended that you use one processor for all your programming exercises but examine the solutions for the other processors and look for points of similarity.

If you will be using 8086 and have not yet read section C.3 in Appendix C, do so now.

7.1 DATA SIZES - POINTS TO REMEMBER

7.1.1 Z80 This is an eight bit processor and most instructions involve eight bit operations. There are, however, some instructions which allow the eight bit working registers to operate in pairs and behave as though they were sixteen bit registers. When data needs to be moved between these register pairs and memory it is moved as two eight bit numbers.

7.1.2 8086 This is a sixteen bit processor but can handle eight and sixteen bit numbers with equal ease. If an instructions refers to a sixteen bit register such as bx the operation will be a sixteen bit operation. If, however, the instruction refers to either the upper or lower nibbles (bh or bl respectively) then an eight bit operation will occur.

7.1.3 68000 The manufacturers data sheet refers to this as a 16-/32- bit processor. It has a sixteen bit data bus but internally the registers are 32 bits wide and instructions are available which allow the processor to operate on either eight, sixteen or thirty two bit data. The size of the data to be used is indicated with suffixes .b, .w, and .l on the operator part of any assembly language instruction e.g.

> **move.b** *num,d0* puts an eight bit number (byte) into the first eight bits of register d0.
>
> **move.w** *num,d0* puts a sixteen bit number (word) into the first sixteen bits of register d0.
>
> **move.l** *num,d0* puts a thirty-two bit number (long word) into the thirty-two bits of register d0. As there is only a sixteen bit data bus the processor must do this in two sixteen bit operations.

7.2 ADDRESSING MODES

There are a number of ways in which a processor might access data and these are referred to as addressing modes. Some manufacturers use different terms for certain addressing modes and represent them differently in assembly language format. These will be indicated where appropriate. As we go through you should remember that the general form for an assembly language instruction is:

> **Z80 and 8086**   *operator*   *destination operand, source operand*
> **68000**   *operator*   *source operand, destination operand*

7.3 IMPLIED ADDRESSING

These are instructions in which the programmer supplies only the operator part of the instruction - the operands are implied by the operator e.g.

> Complement Carry Flag   CCF   (Z80)
>   or   CMC   (8086)

> *(Here the operand is clearly the carry flag)*

7.4 IMMEDIATE ADDRESSING

In this mode of addressing the instruction contains the data to be used. In each of the following examples the source operand is given using immediate addressing.

Z80	`ld`	`a,56h`	loads or puts the hexadecimal value 56 into the accumulator.
8086	`mov`	`al,56h`	moves or puts the hexadecimal value 56 into register al.
68000	`move.b`	`#$56,d0`	moves or puts the hexadecimal value 56 into the first eight bits of register d0.

Note that for Z80 and 8086 immediate data is indicated with just a number but for 68000 immediate data is preceded by the symbol '#'. Note also that hexadecimal values are indicated with the letter 'h' when dealing with Z80 and 8086 but the symbol '$' is used for the 68000.

7.5 IMMEDIATE EXTENDED ADDRESSING

Strictly speaking this term is peculiar to Z80 and is an extension of the immediate addressing mode. It may be used when sixteen bits of immediate data are required e.g.

> `ld`   `bc,1234h` loads the sixteen bit hexadecimal number 1234 into the register pair bc - the value 12 goes into register b and the value 34 into register c.

The data sheets for 68000 and 8086 make no such distinction in the name of the addressing mode. The processors do however have direct equivalents e.g.

```
8086    mov        bx,1234h  moves the hexadecimal value 1234 into register bx.
68000   move.w     #$1234,d5  puts the hexadecimal value 1234 into register d5.
```

7.6 DIRECT/ABSOLUTE ADDRESSING

Most references for the 8086 and Z80 refer to this addressing mode as direct addressing whereas those for the 68000 refer to it as absolute addressing. Essentially they are both the same and in this mode the instruction contains the address of the memory location where the data is to be found or put. In each of the following examples the source operand is given using direct addressing.

Z80 `ld` `a,(7856h)` loads or copies the contents of memory location 7856 hex into the accumulator (a).

8086 `mov` `al,[7856h]` moves or copies the contents of memory location 7856 into register al.

68000 `move.b` `$7856,d0` moves or copies the contents of memory location 7856 into register d0.

Note that *'the contents of'* is indicated by ( ) on the Z80 and by [ ] on the 8086. For the 68000 only the raw number is given. Note also that labels may be used to represent addresses. A section of Z80 assembler code for example might contain:

```
            org         1800h
            . . . . . . . . . . . . . . . . .
            . . . . . . . . . . . . . . . . .
            ld          a,(count)
            . . . . . . . . . . . . . . . . .

            . . . . . . . . . . . . . . . . .
count:                  .byte       23
```

When the assembler first starts work on this program it will use the org statement to establish the start address for the program and then start converting the assembly language into machine code. When it reaches the line containing the instruction `ld a,(count)` it will not know the address of 'count' and so it will leave space for the address to be added. By the time it has completed its first pass it will have made a list of all such addresses (known as a *symbol table*). On its second pass it goes through the program again and uses the symbol table to fill in missing information such as the address of 'count'.

7.7 REGISTER ADDRESSING

The data required will be written to and/or read from an internal register. In the earlier examples access to the destination operand used register addressing e.g.

```
Z80     ld        a,(7856h)
```

The source operand may also use register addressing e.g.

```
Z80     ld        (7856h),a
```

Or indeed both operands may use register addressing e.g.

```
Z80     ld        b,c           loads register b with the contents of register c

8086    mov       bx,cx         moves the contents of register cx into register bx

68000   move.w    d3,d4         moves the contents of the lower 16 bits of register
                                d3 into the lower 16 bits of register d4.
```

Example 7.1

Write a program to add two 8-bit numbers stored in memory locations called *num1* and *num2* and store the result in a third memory location called *num3*.

This is essentially the example we used in Chapter 5 to introduce the use of pseudo code. This time however we are using labels for the addresses and we are now in a position to write and test the entire program. The pseudo code given in chapter 5 was:

> get the first number into the processor.
> get the second number into the processor.
> add the numbers.
> write the result back to memory.

We will need some trial data so assume num1 = 5 and num2 = 3. We need to be sure that num3 does not inadvertently start with the correct answer already in it and so we will initialize num3 to zero. The pseudo code may now be converted to assembly language and added to the skeleton file to produce a working program:

(N.B. Characters in light typeface are from the skeleton file and ***bold italic*** characters are the code to be added.)

```
;***********************
; Ex 7.1 written for Z80 *
;***********************

              .org   1800h

port_a        equ    80h
sw            equ    port_a
led           equ    port_a+1
sw_ctrl       equ    port_a+2
led_ctrl      equ    port_a+3
;****add user equates between these lines ****

;*********************************************
start:        ld     a,0ffh
              out    (sw_ctrl),a
              ld     a,0
              out    (led_ctrl),a

;**** add user data between these lines ****
              ld     a,num1
              ld     b,num2
              add    a,b
              ld     num3,a
              rst    30h

num1          .byte 5
num2          .byte 3
num3          .byte 0
;*****************************************
              .end   start
```

```
***************************
*  Ex 7.1 written for 68000   *
***************************


pit          equ           $800001
sw           equ           pit+$10
sw_ddr       equ           pit+$4
sw_cr        equ           pit+$0c
led          equ           pit+$12
led_ddr      equ           pit+$6
led_cr       equ           pit+$0e

***** add user equates between these lines****

**********************************************
             org           $400400

start        clr.b         sw_cr
             move.b        #$0,sw_ddr
             clr.b         led_cr
             move.b        #$0ff,led_cr

**** add user code between these lines ****
             move.b        num1,d0
             move.b        num2,d1
             add.b         d1,d0
             move.b        d0,num3
             trap          #11
             dc.b          0,0

num1         dc.b          5
num2         dc.b          3
num3         dc.b          0
******************************************

             end           start
```

```
            TITLE           Ex 7.1 written for 8086
            DOSSEG

            .MODEL          SMALL
            .STACK          100h
            .DATA
o_port      dw              0
i_port      dw              0
;**** add user data between these lines ****
num1        db              5
num2        db              3
num3        db              0
********************************************

            .CODE
start:      mov             ax,0
            mov             bx,ds:[408h]
            mov             ax,@data
            mov             ds,ax
            mov             o_port,bx
            inc             bx
            mov             i_port, bx
;**** add user code between these lines ****

            mov             al,[num1]
            mov             bl,[num2]
            add             al,bl
            mov             [num3],al
;*******************************************
            mov             ax,4c00h
            int             21h


            END             start
```

You should enter, assemble and test these programs. For testing it is suggested that at
this stage you single step the program.   If you are using either a Z80 or a 68000 target
system you should refer to the manual to see how this is achieved. If you are using a
PC and do not have a specialist debugger you can use the DOS program *debug.exe*.
The format of the command for invoking and automatically loading a file into debug is:

```
        debug   filename
```

I called my program *ex7_1.asm* and so the program I wanted to single step was called
ex7_1.exe. I therefore invoked debug with the command

```
        debug ex7_1.exe             <cr>
```

If all is well this returns with hyphen as a prompt. To execute a single step type

```
        t                           <cr>
```

Repeat this for each instruction you want to trace. As you trace through your program you should have a copy of your original assembly language program by the side of you so that you can keep track of each instruction. The results of my efforts at this are shown below and I have indicated in bold italics those parts which are of concern to the main part of this program. As you work through you should remember that you have no control as to where the program or data is located and it is highly probable that the register contents on your machine will be different from those shown below.

```
AX=0000  BX=0000  CX=003F  DX=0000  SP=0100  BP=0000  SI=0000  DI=0000
DS=1D0E  ES=1D0E  SS=1D22  CS=1D1E  IP=0013   NV UP EI PL NZ NA PO NC
1D1E:0013 8ED8          MOV     DS,AX
-t
AX=0000  BX=0000  CX=003F  DX=0000  SP=0100  BP=0000  SI=0000  DI=0000
DS=0000  ES=1D0E  SS=1D22  CS=1D1E  IP=0015   NV UP EI PL NZ NA PO NC
1D1E:0015 8B1E0804      MOV     BX,[0408]                    DS:0408=0378
-t
AX=0000  BX=0378  CX=003F  DX=0000  SP=0100  BP=0000  SI=0000  DI=0000
DS=0000  ES=1D0E  SS=1D22  CS=1D1E  IP=0019   NV UP EI PL NZ NA PO NC
1D1E:0019 B8211D        MOV     AX,1D21
-t
AX=1D21  BX=0378  CX=003F  DX=0000  SP=0100  BP=0000  SI=0000  DI=0000
DS=0000  ES=1D0E  SS=1D22  CS=1D1E  IP=001C   NV UP EI PL NZ NA PO NC
1D1E:001C 8ED8          MOV     DS,AX
-t
AX=1D21  BX=0378  CX=003F  DX=0000  SP=0100  BP=0000  SI=0000  DI=0000
DS=1D21  ES=1D0E  SS=1D22  CS=1D1E  IP=001E   NV UP EI PL NZ NA PO NC
1D1E:001E 891E0800      MOV     [0008],BX
DS:0008=0000
-t
AX=1D21  BX=0378  CX=003F  DX=0000  SP=0100  BP=0000  SI=0000  DI=0000
DS=1D21  ES=1D0E  SS=1D22  CS=1D1E  IP=0022   NV UP EI PL NZ NA PO NC
1D1E:0022 43            INC     BX
-t
AX=1D21  BX=0379  CX=003F  DX=0000  SP=0100  BP=0000  SI=0000  DI=0000
DS=1D21  ES=1D0E  SS=1D22  CS=1D1E  IP=0023   NV UP EI PL NZ NA PO NC
1D1E:0023 891E0A00      MOV     [000A],BX
DS:000A=0000
-t
AX=1D21  BX=0379  CX=003F  DX=0000  SP=0100  BP=0000  SI=0000  DI=0000
DS=1D21  ES=1D0E  SS=1D22  CS=1D1E  IP=0027   NV UP EI PL NZ NA PO NC
1D1E:0027 A00C00        MOV     AL,[000C]                    DS:000C=05
-t
AX=1D05  BX=0379  CX=003F  DX=0000  SP=0100  BP=0000  SI=0000  DI=0000
DS=1D21  ES=1D0E  SS=1D22  CS=1D1E  IP=002A   NV UP EI PL NZ NA PO NC
1D1E:002A 8A1E0D00      MOV     BL,[000D]                    DS:000D=03
-t
AX=1D05  BX=0303  CX=003F  DX=0000  SP=0100  BP=0000  SI=0000  DI=0000
DS=1D21  ES=1D0E  SS=1D22  CS=1D1E  IP=002E   NV UP EI PL NZ NA PO NC
1D1E:002E 02C3          ADD     AL,BL
```

```
-t
AX=1D08  BX=0303  CX=003F  DX=0000  SP=0100  BP=0000  SI=0000  DI=0000
DS=1D21  ES=1D0E  SS=1D22  CS=1D1E  IP=0030   NV UP EI PL NZ NA PO NC
1D1E:0030 A20E00        MOV     [000E],AL                       DS:000E=00
-t
AX=1D08  BX=0303  CX=003F  DX=0000  SP=0100  BP=0000  SI=0000  DI=0000
DS=1D21  ES=1D0E  SS=1D22  CS=1D1E  IP=0033   NV UP EI PL NZ NA PO NC
1D1E:0033 B8004C         MOV     AX,4C00
-t
AX=4C00  BX=0303  CX=003F  DX=0000  SP=0100  BP=0000  SI=0000  DI=0000
DS=1D21  ES=1D0E  SS=1D22  CS=1D1E  IP=0036   NV UP EI PL NZ NA PO NC
1D1E:0036 CD21           INT     21   ;Terminate a Process
```

Notice that the labels **num1, num2** and **num3** have been translated into offsets **000C, 000D** and **000E** (these may be different on your machine).   These offsets are for data and so they are relative to the data segment register. To display memory contents on the screen the format is:

> **d**segement address:offset

Looking at the output above it can be seen that, when dealing with the main part of this program, the data segment register (DS) has a value 1D21. The first memory location affected by our program is at offset 000C and so, to see the effects of this program, it is necessary to type

> d1d21:c   (or d1d21:000c).

The results were:

```
1D21:0000                                        05 03 08 FA
...z
1D21:0010  CB A1 6A 2C 40 74 13 48-E8 55 FF 3D 24 00 74 05
K!j,@t.HhU.=$.t.
1D21:0020  3D 75 00 75 05 83 06 6A-2C 02 CB 35 00 34 00 23
=u.u...j,.K5.4.#
```

and you can see where each item of data in the program was placed.

If you examine the instruction sets for the 8086 and the 68000 you will see that the add instruction can have one of its operands in memory. Thus the main part of the program could have been

8086

```
        mov    al,[num1]    instead of    mov    al,[num1]
        add    al,[num2]                  mov    bl,[num2]
        mov    [num3],al                  add    al,bl
                                          mov    [num3],al
```

68000

```
move.b    num1,d0    instead of    move.b    num1,d0
add.b     num2,d0                  move.b    num2,d1
move.b    d0,num3                  add.b     d1,d0
                                   move.b    d0,num3
```

(If you are using either 8086 or 68000 try this variant of the program.)

Now complete exercise 7.1 at the end of this chapter.

7.8 ADDRESSING I/O

68000. The 68000 treats I/O devices in exactly the same way as any memory device. This means that any addressing mode used to access memory may also be used to access an I/O device. The only restrictions come from the hardware used to construct an I/O port. If, for example, D'type latches have been used to construct an output port then, if we attempt to read from that address, the result will be meaningless.

Z80. In section 4.11 you saw that it was possible to design a Z80 based system so that I/O devices could be accessed either by normal memory transfer instructions (known as **memory mapped I/O**) or they could be accessed with special I/O instructions (known as **I/O mapped I/O**). For the rest of this book we will assume that the system you are using has I/O mapped I/O. Also, although there are a number of addressing modes that may be used, this book will only use the following instructions

```
        in        a,(n)
e.g.    in        a,(80h)
```
reads contents of port at I/O address (80) and copies the contents into the accumulator.

```
        out       (n),a
c.g.    out       (81h),n
address
```
send contents of accumulator out to port at I/O address (81h)

8086. 8086 based systems, like Z80 based systems, may be designed with memory mapped I/O but this is rarely done. I/O mapped I/O is usually adopted. If you are using the 8086 processor for your programming exercises, this book assumes that you will be using the printer book for I/O and that any programs you write will be an expansion of the 8086 skeleton file given in section 6.4. If this is the case you may read the input port with the instructions:

```
mov       dx,[i_port]
in        al,dx
```

and write to the output port with the instructions

```
mov       dx,[o_port]
out       dx,al
```

Now complete exercise 7.2 at the end of this chapter.

7.9 LOGICAL OPERATIONS

In this context 'logical' refers to the Boolean operations *AND, OR* and *exclusive-OR.*
Consider a system which has two inputs A and B and a single output F.

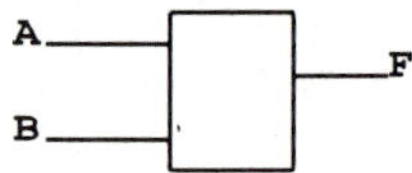

If the system performs an **AND** operation the output F will be logical 1 if, and only if,
both inputs are logical 1. If the system performs an **OR** operation the output will be a
logical 1 if any or all of its inputs are logical 1. If the system performs an **exclusive-
OR** operation (usually referred to as *XOR* in instruction sets) the output will be logical
1 if either of its two inputs are logical 1 but not both.

When a microprocessor performs one of these logical operations it does so on a bit
by bit basis e.g.

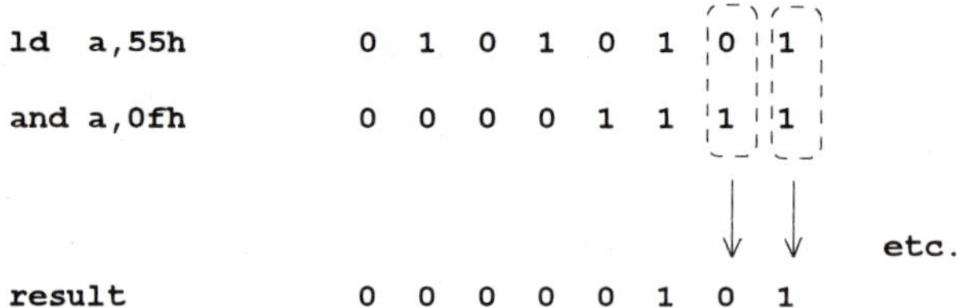

```
ld   a,55h        0  1  0  1  0  1  0  1

and a,0fh         0  0  0  0  1  1  1  1

                                          etc.

result            0  0  0  0  0  1  0  1
```

When writing programs for control applications it is often necessary to manipulate bits
within bytes. The logical operators may be used for this purpose. If we think of the
gate above slightly differently:

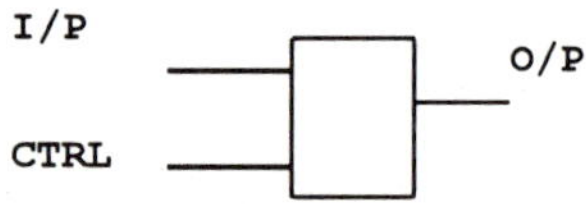

we have:

CTRL	I/P		O/P		
		AND	OR	XOR	
0	0	0	0	0
0	1	0	1	1
1	0	0	1	1
1	1	1	1	0

If CTRL = 0	AND causes	O/P always 0
	OR causes	O/P = I/P
	XOR causes	O/P = I/P

If CTRL = 1	AND causes	O/P = I/P
	OR causes	O/P always 1
	XOR causes	O/P = inverse of I/P

Example 7.2

A microprocessor based system has an 8 bit input port and an 8 bit output port. For development purposes the inputs are connected to switches and the outputs to LEDs. Write the pseudo-code for a section of program which will display the state of the switch at bit 6 only on the LEDs - all other LEDs are to be extinguished.

We can force bits to be zero by ANDing with logical 0 and preserve bits by ANDing with logical 1. Thus we are looking for:

```
8 bits from switches        xxxx xxxx
40 hex                      0100 0000
                            _________
result of ANDing            0x00 0000
```

and the corresponding pseudo-code is:

        get the value on the switches
        logically AND with 40 hex
        display the result on the LEDs

Isolating a bit or bits by using an AND operation is often refered to as *masking* or *bit masking*.

Now complete exercises 7.3, 7.4, 7.5 and 7.6 at the end of this chapter.

In the following exercises num1, num2, num3 and num4 are data items in consecutive memory locations and should take the values num1 = 3, num2 = 7, num3 = 9 and num4 = 0.

Exercise 7.1
Write and test a program which will add num1, num2 and num3 and leave the result in num4.

Exercise 7.2
Amend the program from Exercise 7.1 so that the result is sent directly to the LEDs instead of num4.

Exercise 7.3
Examine the instruction set for the processor on which you are working and, for each of the logical operators, write down:

  a)   The assembly language format for each of the instructions.
  b)   The addressing modes allowed for the source operand.
  c)   The addressing modes allowed for the destination operand.

Exercise 7.4
Write and test a program which will OR num1 with num2 and then AND the result with num3. The result is to be displayed on the LEDs.

Exercise 7.5
Write and test a program which will combine the 4 least significant bits of num1 with the 4 most significant bits from the switches and display the results on the LEDs.

Exercise 7.6   *** suggested for portfolio ***
Write a program to read the switches and dispaly the state of bits 5 and 6 only on the LEDs. All other LEDs are to be extinguished.

8

Assembly Language Programming - 2

In chapter 2 it was mentioned that, when the ALU performs an operation, certain information relating to the result is stored in the flags register. This section adds more detail and shows how the flags may be used to determine which instructions are executed next.

8.1 ALU AND THE FLAGS
8.1.1 The Problem
Consider the following examples written in Z80 assembler:

```
         (a)                                    (b)

ld   a,81h      1000 0001        ld   a,41h      0100 0001
add  a,80h      1000 0000        add  a,40h      0100 0000
                ___________                      ___________
              1 0000 0001                        1000 0001
```

In (a) we are attempting to add two eight bit numbers to produce a nine bit answer. The instruction **add a,80h** tells us that the result is placed in the accumulator but the Z80 only has an eight bit accumulator. The Z80 responds to this by putting **0000 0001** in the accumulator and making note that a *carry* has occurred by setting a bit in the flags register to logical 1.

In eight bit signed arithmetic bit 7 represents the sign of the number. In example (b) we are apparently adding two positive numbers (bit 7 = 0 in both cases) and getting a negative result (bit 7 of the result = 1) i.e. the sign bit has been corrupted. This is called *overflow*. Again the processor will respond to this by setting a bit in the flags register to logical 1.

*Whenever such anomalies occur it is the **responsibility of the programmer** to take appropriate action.*

8.1.2 Flag and Flag Setting

The structure of the flags register or status register varies from one processor to another. For the processors being considered by this book the structures are:

Z80

												S	Z		H		P/V	N	C

8086

						OF	DF	IF	TF	SF	ZF		AF		PF		CF

68000

T		S			I2	I1	I0				X	N	Z	V	C

Whether or not a flag is affected will be determined by:

    a)   the value of the result.
    b)   whether or not the last instruction executed will affect flags.
    c)   the size of the operation and the size of the result.

Item c) may seem a little odd. Earlier we saw that, for the Z80, loading the accumulator with 81 hex and adding 80 hex caused a carry. The equivalents for this on a 68000 and a 8086 are respectively:

```
move.b   #$81,d0              mov     al,81h
add.b    #$80,d0              add     al,80h
```

In both cases we are using eight bit data and expecting an eight bit result. Thus the generation of a nine bit result causes the carry flag to be set. If however these instructions had been:

```
move.w   #$81,d0              mov     ax,81h
add.w    #$80,d0              add     ax,80h
```

Then we would be using sixteen bit data and expecting a sixteen bit result and so the carry flag would not be set. For the programming examples in this book we will be concerned with recognizing the following conditions:

Result Zero This is indicated by the **Z** flag on the Z80 and 68000 and by the **ZF** flag on the 8086. It is set to a logical 1 if the result of an operation has a result of zero otherwise it is reset to logical 0. Remember that the size of the data is important. If, for example you are doing an **eight bit operation** then 80h + 80h will cause the Z flag to be set because the **first eight bits** (i.e. the least significant eight bits) of the result are zero.

Carry This is indicated by the **C** flag on the Z80 and 68000 and by the **CF** flag on the 8086. The carry flag will be set if the size of the result is bigger than the size of the data permitted by the operation e.g. the carry flag will be set if an eight bit operation produces a nine bit answer or if a sixteen bit operation produces a seventeen bit answer. The carry flag will also be set by a subtract operation if there is a need to borrow from a more significant bit e.g. 80h - 81h. Carry must not be confused with **overflow** which will be dealt with shortly.

Sign   Most processors automatically regard the most significant bit of a result as indicating its sign. This happens whether or not your are intentionally using signed numbers. The sign is indicated by the **S** flag on the Z80, the **N** flag on the 68000 and the **SF** flag on the 8086.

Overflow This is indicated by the **P/V** flag on the Z80, the **V** flag on the 68000 and by the **OF** flag on the 8086. The term overflow is unfortunate since this name often causes it to be confused with carry. **Overflow means that the sign bit has been corrupted**. Consider the following examples of eight bit addition:

```
          (a)                      (b)                      (c)
      0100  0001              1000  0001              1100  0001
      0100  0010              1000  0010              1100  0010
      ───────────            ─────────────            ─────────────
      1000  0011             1 0000  0011             1 1000  0011
carry     1                     1                       1   1
```

In (a) we add two positive numbers (bit 7 = 0) and get an indicated negative result (bit 7 = 1). In (b) we add two negative numbers (bit 7 = 1) and get an indicated positive result (bit 7 = 0). In (c) we add two negative numbers (bit7 = 1) and get an indicated negative result (bit 7 = 1).   Note that the only example with a correct sign bit is (c) and this had a carry from bit 6 to bit 7 and a carry from bit 7 to bit 8. In general, for eight bit arithmetic, we can say that the sign bit will be corrupted if there is a carry from bit 6 to bit 7 **or** from bit 7 to bit 8 **but not** if there is a carry from **both**. Similarly, with sixteen bit arithmetic, the sign bit will be corrupted if there is a carry from bit 14 to bit 15 or from bit 15 to bit 16 but not from both.

The symbol P/V used on the Z80 indicates that, on this processor, it is a dual purpose flag. If the processor has just executed an arithmetic operation the flag becomes a V flag and indicates overflow. However, if it has just executed a logical operation (AND, OR, XOR), the flag becomes a P flag indicating the parity of the result.

Extended The extended or **X** flag is peculiar to the 68000 and is used as a carry for multiple precision operations. It is also used with shift and rotate operations.

There will be occasions when it becomes necessary to put the carry flag in a particular condition before commencing an operation. The Z80 and 80x86 processors have instructions for doing this:

| set carry to logical 1 | **SCF** (Z80) | **STC** (80x86) |
| clear carry to logical 0 | **CLC** | |

To clear the carry flag on the Z80 it is necessary to first set the flag with **SCF** and then complement it (invert it) with **CCF**.

On the 68000 it is possible to alter the conditions of most of the flags by using the logical operators **ORI** and **ANDI** in a masking technique to set or clear individual bits in the status register.

e.g.   `ORI.W  #$1,SR`   sets bit 0 of the status register (the carry flag) to a logical 1

   `ANDI.W #$FFFE,SR`clears bit 4 of the status register (the extend flag X) to a logical 0.

The lower eight bits of the status register are also known as the condition code register and so, using byte size operands, the instructions above could also be written as :

   `ORI.B  #$1,CCR` and `ANDI.B #$FE,CCR`

Now complete Exercise 8.1 at the end of this chapter

8.2 JUMP AND BRANCH INSTRUCTIONS

Unless instructed to do otherwise a microprocessor will execute instructions in strict order. Jump and branch instructions may be used to cause the processor to break out of its current sequence of instructions and start executing a different sequence of instructions. Jump and branch instructions may be either **conditional** or **unconditional**. Unconditional instructions will be executed whenever the instruction is encountered. However, if a conditional jump or branch instruction is encountered, the processor will look to the flags register to see if the conditions for a jump have occurred. Some examples of jump instructions are given below:

Instruction	flag tested	Assembly language format		
		Z80	**68000**	**8086**
Always jump *label* or branch	none	jp *label*	bra *label*	jmp
Jump (branch) if *label* carry set	C or CF	jp c,*label*	bcs *label*	jc
Jump (branch) if *label* carry clear	C or CF	jp nc,*label*	bcc *label*	jnc

Jump (branch) if equal to zero	Z or ZF	jp z,*label*	beq *label*	je *label* (or jz *label*)
Jump (branch) if not equal to zero	Z or ZF	jp nz,*label*	bne *label*	jne *label* (or jnz *label*)

Other conditional jump and branch instructions exist and you should check the instruction set for the processor you are using for details. At this point it is worth noting that both the 8086 and the 68000 have jump instructions which can test more than one flag

e.g.	8086	ja *label*	means jump if above and this tests both the CF and ZF flags.
	68000	bgt *label*	means branch if greater and this tests both the C and Z flags.

8.3 THE COMPARE INSTRUCTION

The compare instruction performs a subtract operation but does not store the answer anywhere. Its sole function is to condition flags. Suppose that it is necessary to keep reading a sensor until a reading of 54 hex is detected. The assembly language for this might be:

```
          Z80                      8086                    68000

read:  in a,(sensor)        in   al,sensor        move.b $sensor,d0
       cp 54h               cmp  al,54h            cmp.b  #$54,d0
       jp nz,read           jne  read              bne    read
```

In each case the first instruction reads the contents of the sensor into an internal register. The compare instruction determines the result of subtracting 54 hex from this register and, if the result is zero, the Z flag (ZF flag) is set.

The last instruction looks at the Z flag to see if a jump should occur. If the result of the subtraction was not zero, then the sensor did not have a value of 54 and the Z flag will not have been set. The **not zero** or **not equal** condition is said to be *true* and so the program counter will be adjusted so as to make the processor jump back to read the sensor again i.e. the jump is taken.

If the value on the sensor is 54 hex then the result of the subtraction will be zero, the Z flag will be set and the **not zero** or **not equal** condition is said to be *false*. This means that the jump will not be taken and the processor will continue with whatever follows the jump instruction.

Example 8.1

Write a program which will read two 8 bit unsigned numbers from memory locations num1 and num2, and display the larger of the two values on the LEDs.

The pseudo code for the main part of this program might be:

```
        get num1
        get num2
        compare numbers
        if  num1 > num2
                display num1
                go to exit routine
    else
                display num2

        execute exit routine
```

Remember that the compare instruction performs a subtract operation. If num1 - num2 is positive then num 1 is the greater. To determine if a result is positive or negative we must examine the carry flag. **Positive** results are indicated by **carry clear** and **negative** results are indicated by **carry set.** Thus, if we use num1 = 9 and num2 = 5 as trial data, the following code needs to be added to the skeleton files:

Z80

```
; * * * * * * * * * * * * * * * *  Enter user code between these lines  * * * * *

        ld      a,num1
        ld      b,num2
        cp      b               ; evaluates a - b which in this
                                ; case is num1 - num2

        jp      c,smaller ; jump if num2 smaller
        out     (led),a   ; num1 already in accumulator
        jp      exit

smaller: ld     a,b             ; move num2 into accumulator
        out     (led),a   ; send num2 to leds

exit:   rst     30h             ; use whatever

num1:   .byte   9
num2:   .byte   5
```

8086

```
; * * * * * * * * * * * * * * * *  enter user data between these lines  * * * * *


num1        db        9
num2        db        5
; * * * * * * * * * * * * * * * * * * * * * * * * * * * * * * * * * * * * * *


; * * * * * * * * * * * * * * * *  enter user code between these lines  * * * * *

            mov       al,num1
            mov       bl,num2
            cmp       al,bl        ; evaluates al - bl which in this
                                   ; case is num1 - num2

            jc        smaller   ; jump if num2 smaller
            mov       dx,[o_port]
            out       dx,al        ; num1 already in accumulator
            jp        exit

smaller:    mov       dx,[o_port]
            out       dx,bl       ; num2 already in bl

; * * * * * * * * * * * * * * * * * * * * * * * * * * * * * * * * * * * * *
exit:       mov    ax,4c00h etc.
```

68000

```
* * * * * * * * * * * * * * *  Enter user code between these lines  * * * * *
            move.b $num1,d0
            move.b $num2,d1
            cmp.b  d1,d2        ; evaluates a - b which in this
                               ; case is num1 - num2

            bcs       smaller   ; jump if num2 smaller
            move.b d0,$led    ; num1 already in d0
            jp        exit

smaller:    move.b d1,$led    ; num2 already in d1

num1:       dc.b    9
num2:       dc.b    5
* * * * * * * * * * * * * * * * * * * * * * * * * * * * * * * * * * * * * * *
```

Now complete exercises 8.2, 8.3 and 8.4 at the end of this chapter

8.4 REGISTER INDIRECT ADDRESSING

In this mode of addressing the address of the memory location to be accessed is held in one of the processor's registers:

e.g.1   Z80   ld   a,(hl)

This instruction may verbalized as *the contents of* **the address pointed to by hl** *are loaded into the accumulator.*   The term **pointed to** means, in this case, that the contents of register pair hl are placed on the address bus instead of either the program counter or the memory address register. Many Z80 instructions allow this form of addressing with register pairs BC and DE as well as HL.

e.g.2   68000   move.b   (A0),D3

This means move the contents of the address pointed to by address register A0 into data register D3. Address registers A0 through to A7 may be used for this purpose but A7 should be avoided by the novice programmer since it is used for other purposes. To make tabulated data easier to access the 68000 has two variants of this addressing mode:

> **Postincrement** where the address register is incremented as soon as it has been used. The syntax is *(An)+.*
> **Predecrement** where the address register is decremented immediately before it is used. The syntax is *-(An).*

e.g.3   8086   mov   ax,[bx]

This means move the contents of the address pointed to by bx into register ax. Registers BX, BP, SI and DI may be used for this purpose. Note that these are all 16 bit registers and the 8086 has a 20 bit address bus. By default the address accessed by the instruction above will be relative to the data segment register. The reader should be aware that more advanced forms of this addressing mode are also available but will not be covered in this book.

Example 8.2
Write a program using register indirect addressing to add together four 8 bit numbers in adjacent memory locations and store the result immediately after the fourth number.

*To indicate the **"address pointed to"** in the pseudo code I will use the symbol **[pointer]***

As a first try the pseudo code might be:

```
clear a register for the total
load the address of the first number into a register     (this is the pointer)
total = total + [pointer]
increment pointer
total = total + [pointer]
increment pointer
total = total + [pointer]
increment pointer
total = total + [pointer]
increment pointer
store total at [pointer]
```

This would be all right provided there were not too many numbers to be added. For larger quantities of data it would be better to use a loop. When adding four items of data in a loop the pseudo code might be:

```
clear a register for the total
set  loop counter to 4
load the address of the first number into a register
next:
        total = total + [pointer]
        increment pointer
        decrement loop counter
        if  loop counter not zero go to next
store total at [pointer]
```

Translating this into assembly language we have:

Z80

```
;********** add user code between these lines**********
        ld      a,0             ;clear register for total
        ld      b,4             ;loop counter = 4
        ld      hl,num          ;point hl at data

next:   add     a,(hl)          ;tot = tot + (pointer)
        inc     hl              ;increment pointer
        dec     b               ;decrement counter
        jp      nz,next         ;jump if not zero

        ld      (hl),a          ;store total
        rst     30h

num:    .byte   1,2,3,4,0
```

```
;*****************************************************
68000
********* add user code between these lines***********

        move.b   #0,D0            *clear register for total
        move.b   #4,D1            *loop counter = 4
        movea.l  #num,A0          *point A0 at data

next    add.b    (A0)+,d0         *tot = tot + (pointer)
                                  *(A0)+ increments pointer
        sub.b    #1,D1            *decrement counter
        bne      next             *jump if not zero
        move.b   D0,(A0)          *store total

        TRAP     #11
        DC.W     0


num     dc.b     1,2,3,4,0
*****************************************************
```

Note the use of the instruction:

```
movea.l   #num,A0
```

When trying to load data into a 68000 address register it is important to use a **movea** (**move address**) instruction.


```
8086
;************* enter user data between these lines ******

nums    db       1, 2, 3, 4, 0


;*****************************************************


;************* enter user code between these lines
        mov      ax,0             ;clear register for total
        mov      cx,4             ;set loop counter = 4
        mov      bx,OFFSET nums   ;point bx at data

next:   add      al,[bx]          ;tot = tot + (pointer)
        inc      bx               ;increment pointer
        dec      cx               ;decrement loop counter
        jne      next             ;jump if counter not zero
        mov      [bx],al          ;store total
```

Note the use of the instruction:

```
    mov        bx,OFFSET nums
```

The command **OFFSET** forces the assembler to calculate the address relative to the data segment register.

Now complete Exercises 8.5 and 8.6 at the end of this chapter.

Exercise 8.1
Predict the state of the carry, overflow, zero and sign flags when the following operations are executed:

    (a) $51_{16} + 31_{16}$ (b) $D3_{16} + 39_{16}$ (c) $80_{16} + 80_{16}$

    (d) $AA_{16}\ OR\ 0f_{16}$ (d) $5A_{16}\ XOR\ 5A_{16}$

Exercise 8.2
Write a program which will read two 8 bit numbers in the range 00 hex to 0f hex from memory locations num1 and num2, and display the smaller of the two values on the LEDs.

Exercise 8.3
Rewrite Exercise 8.2 so that, if num1 = num2, all leds are illuminated.

Exercise 8.4
A check sum is to be placed at the end of a block of four bytes of data. The value of the check sum is chosen so that, when it is added to the total of the four preceding bytes, the result will be zero when any ninth bit carry is dropped (e.g. If the total of the four bytes was FA_{16} the check sum would be 06_{16} so that $FA_{16} + 6_{16} = 100_{16}$ and this is 00_{16} when the ninth bit carry is dropped). Write a program to calculate the value of the check sum and place it in the correct memory location.

Exercise 8.5
Four numbers in the range 0 to 9 are held in consecutive memory locations. Before they can be sent to a printer it is necessary to convert them to their ASCII equivalent. Write a program to achieve this using register indirect addressing when appropriate.

Exercise 8.6
Refer to Case Study 1. Write a program to implement the first paragraph of routine (a). The program should be written as a stand alone item and you should ignore the initial need for the switches to be at 00.

Exercise 8.7   - *** *suggested for portfolio* ***
Refer to Case Study 2.   Circuit 2 puts a logical 1 onto bit 6 of the input port when the
switches are closed and the master switch puts a logical 1 onto bit 5 of the input port
when the alarm system is armed. The siren is held off by a logical 1 on bit 7 of the
output port. Write a program which will send the appropriate signal to the siren if the
system is armed and the input from circuit 2 changes to logical 0.

9

Program Flow Control

In Chapter 5 you learnt that **JSP** permits three control structures only: **sequence, selection and iteration.** You have been dealing with sequence in all of the programming examples covered so far: it is simply the execution of instructions in the order in which they occur. The early part of this chapter examines methods of implementing the selection and iteration control structures, both of which require data or results to be tested.

For a more complete implementation of JSP, division into subprograms is required. At assembly code level this means using **subroutines.** The later parts of this chapter looks at subroutines and a very important related item - the **stack.**

9.1 TESTING FOR PROGRAM FLOW CONTROL

In the context of program flow control a test can have one of two results only: it is either **true** or it is **false.** Earlier books on programming would show a test on a flow chart as:

In machine code or assembly language a test consists of two instructions: **an instruction which can set flags** and a **conditional jump or branch instruction.** Some examples of these are shown below:

Z80		68000		8086	
dec	b	sub.b	#1,D0	dec	ax
jp	nz,next	bne	next	jne	next
and	80h	and.b	#$80,D2	and	bh,80h
jp	z,clear	beq	clear	je	clear
cp	54h	cmp.b	#$54,D1	cmp	al,54h
jp	c,neg	bcs	neg	jc	neg

The first set of examples involving *dec* and *sub* use an arithmetical operation to set flags and illustrate a test for zero. In each of these cases the branch or jump is taken if the result is ***not zero***.

The second set of examples use a logical operation to set flags and illustrates the use of masking. In these examples the *AND* operation preserves bit 7 and sets all other bits to zero. This means that the final result will be either zero or not zero depending on the state of bit 7. The jump or branch is taken if the result is zero.

Although the compare instruction does not write the result anywhere it is the arithmetical operation of subtraction and will set flags in the same way as other subtraction operations. The final set of examples show how the carry flag may be used to test the sign of the result after a compare instruction has be executed. If the carry flag is set the result is negative. If the carry flag is clear the result is either positive or zero. (You may remember from chapter 8 that the sign flag is unreliable for simple sign tests since it may be corrupted by arithmetical operations.)

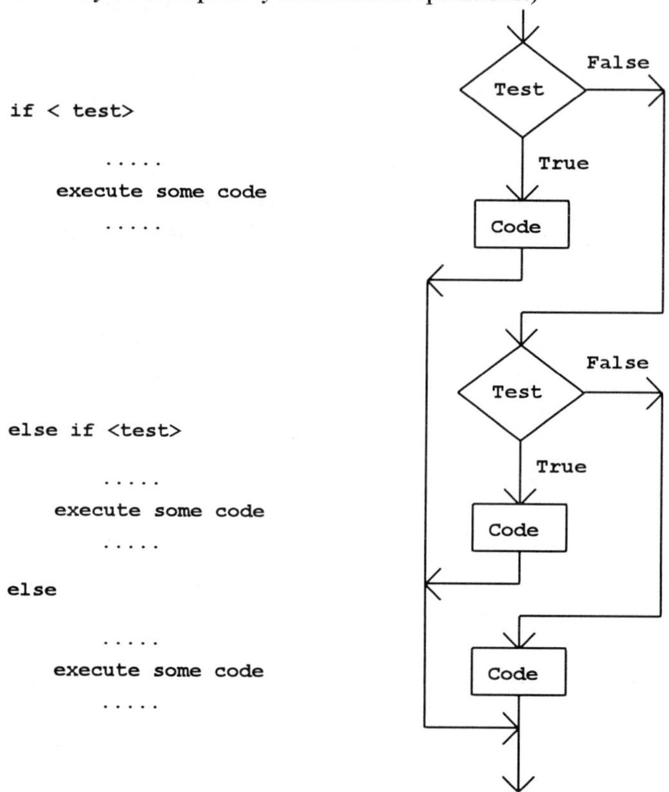

Figure 9.1

9.2 SELECTION

Selection in JSP requires that one, and only one, of a number of sections of code be executed. The pseudo code and flow chart equivalents of this are shown in Figure 9.1.

Example 9.1
Write a program which will read the switches. If the value on the switches is greater than 80 hex bit 7 on the LEDs is to be illuminated. If the value is between 41 hex and 80 hex bit 6 on the LEDs is to be illuminated. If none of these conditions apply all LEDs are to be extinguished.

The pseudo code might be:

```
        read switches
        if switches > 80 hex
                display 80 hex
        else if switches > 40 hex
                display 40 hex
        else
                display zero
```

and the corresponding additions to the skeleton files are:

Z80

```
;*** add user code between these lines ******************
                in      a,sw            ;read switches
                cp      80h             ;check if > 80
                jp      c,try_40h       ;jump if not > 80
                ld      a,80h           ;code to display 80
                out     (led),a         ;         "
                jp      exit            ;

try_40h:        cp      40h             ;check if > 40
                jp      c,all_out       ;jump if not > 40
                ld      a,40h           ;code to display 40
                out     (led),a         ;         "
                jp      exit            ;jump to exit routine

all_out:        ld      a,0h            ;code to display 00
                out     (led),a         ;         "

;************************************************************
exit:           rst     30h
```

8086

```
;** enter user data between these lines ******************
            mov       dx,[i_port]
            in        al,dx           ;read switches
            cmp       al,80h          ;check if > 80
            jc        test2           ;jump if not > 80
            mov       al,80h          ;code to display 80
            mov       dx,[o_port] ;           "
            out       dx,al           ;           "
            jmp       exit            ;jump to exit routine

test2:      cmp       al,40h          ;check if > 40
            jc        display_0       ;jump if not > 40
            mov       al,40h          ;code to display 40
            mov       dx,[o_port] ;           "
            out       dx,al           ;           "
            jmp       exit            ;jump to exit routine

display_0:  mov       al,0            ;code to display 00
            mov       dx,[o_port] ;           "
            out       dx,al           ;           "
;****************************************************************
```

68000

```
************   user code goes between these lines *********
            move.b   sw,D0
            cmp.b    #$80,D0
            bcs      try_40
            move.b   #$80,led
            jmp      exit

try_40      cmp      #$40,D0
            bcs      all_out
            move.b   #$40,led
            jmp      exit

all_out     move.b   #$0,led
*****************************************************************
```

9.3 ITERATION

This is when a section of code needs to be repeated a number of times. The code is put into a loop for which the pseudo code and flow chart equivalents are shown in Figure 9.2.

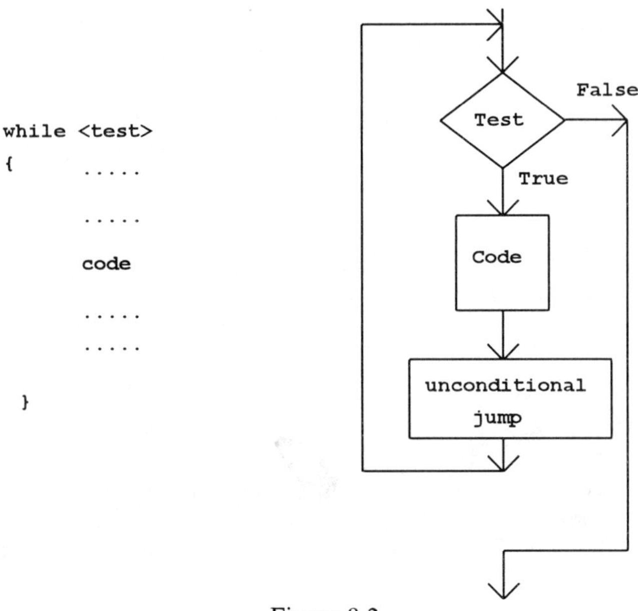

Figure 9.2

Example 9.2

Write a program which will read the switches and display the contents on the LEDs until a value less than 20 hex is read. The program shall then extinguish all LEDs and stop.

The pseudo code might be:

```
read switches
while value >= 20 hex
{
        copy switches to LEDs
}
send 0 hex to LEDs
end.
```

and the corresponding additions to the skeleton files are:

Z80

```
;********** enter user code between these lines **********
next:           in      a,(sw)          ;read switches
                cp      20h             ;check if < 20
                jp      c,display_0     ;jump if result negative
                out     (led),a         ;send to leds
                jp      next            ;end of loop, go back for
                                        ;next reading


display_0       ld      a,0
                out     (led),a         ;send zero to leds
;**************************************************************************
```

8086

```
;********** enter user code between these lines **********
next:           mov     dx,[i_port]
                in      al,dx           ;read switches
                cmp     al,20h          ;check if < 20h
                jc      display_0       ;jump if result negative
                mov     dx,[o_port]
                out     dx,al           ;send to leds
                jmp     next            ;end of loop, go back for
                                        ;next reading


display_0:      mov     al,0
                mov     dx,[o_port]     ;send zero to leds
                out     dx,al
;**************************************************************************
```

68000

```
********** enter user code between these lines **********
next            move.b  sw,D0           ;read switches
                cmp.b   #$20,D0         ;check if < 20h
                bcs     display_0       ;jumpr if result negative
                move.b  D0,led          ;send to leds
                bra     next            :end of loop, go back for
                                        ;next reading


display_0       move.b  #0,led          ;send zero to leds
;**************************************************************************
```

Now complete exercises 9.1 and 9.2 at the end of this chapter.

9.4 SUBROUTINES

A subroutine is a section of code which may be called from anywhere in the program and, once completed, will automatically return to the part of the program from which it was called. The operation of a subroutine is shown in Figure 9.3:

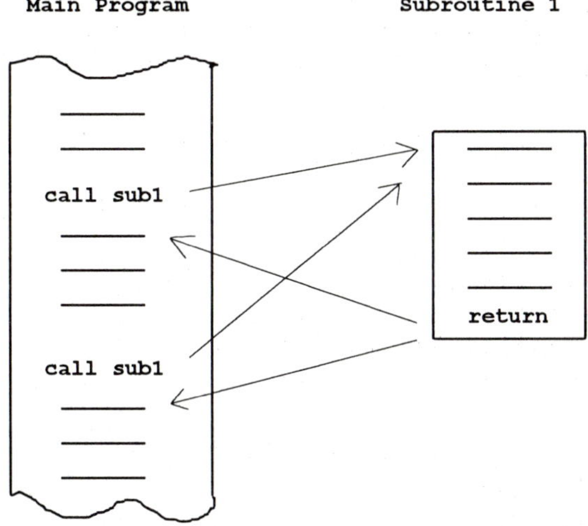

Figure 9.3

Quite often it will be necessary for one subroutine to call another. Subroutines which do this are known as **nested subroutines** and the arrangement is illustrated in Figure 9.4

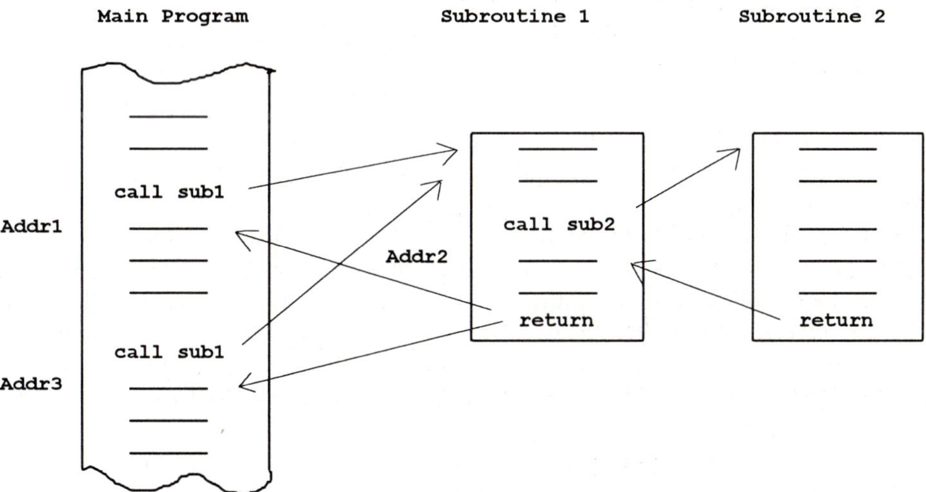

Figure 9.4

The main reasons for choosing to use a subroutine are:

a)   there is a section of code which is needed by a number of parts of the program. Using a subroutine means that the code need only be entered once.

b)   you are rigorously applying a structured technique such as JSP which demands the use of a subroutine.

The main advantages of subroutines are:

a)   shorter object code (assembly code).
b)   improved program structure.
c)   improved program readability.

The main disadvantages of subroutines are:

a)   increased running time since additional instructions are required to jump to, and return from, each subroutine.
b)   corruption of register contents.

(If you are familiar with subroutines in high languages then you may be surprised by the last item. High level language compilers automatically preserve register content for you. However, at assembler level, it is the ***programmers responsibility*** to ensure that registers are not corrupted.)

9.4.1 Calling a Subroutine

You may recall from the discussion of the **fetch-execute cycle** in Chapter 2 that the program counter is incremented as soon as it has been used. This means that, once an instruction has been read in to the processor, the program counter holds the address of the first byte of the next instruction. When an ordinary jump or branch instruction is executed the contents of the program counter are over written with the destination address for the jump.

When a subroutine is called a similar process takes place. However, for subroutines, it must be possible for an automatic return to the main program to be made. To do this the processor must make a note of the return address. The way this is achieved varies from one processor to another. Most processors (but by no means all) save the return address in an area of **RAM** known as the **stack**. In general any area of RAM may be used but it is usual to put user programs in the lower addresses and the stack at the higher addresses.

If you refer back to Figure 9.4 you will see that there may well be a need to store more than one return address. The processor keeps track of where the last address is stored in a register which is generally referred to as the **stack pointer** or **SP** (on the 68000 address register A7 is used as the stack pointer). Depending upon the processor being used, the stack pointer will contain the address of *either* the last useful item on the stack *or* the next available space on the stack. The diagrams which follow illustrate how the stack and stack pointer would be affected by a program structure shown in Fig. 9.4 for a processor whose stack pointer points to the last useful item on the stack.

`1. At start of main program`

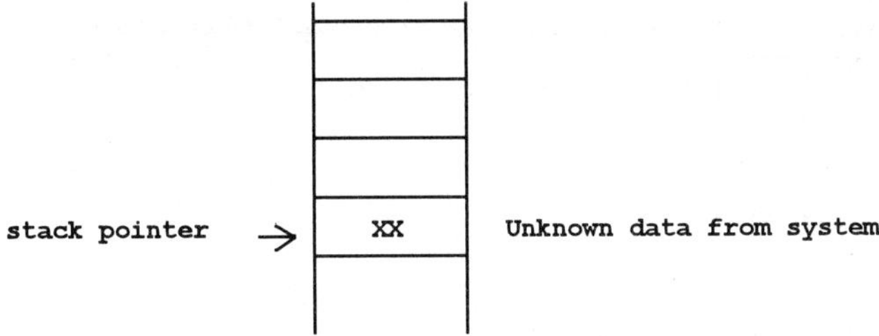

`2. After first call to sub 1`

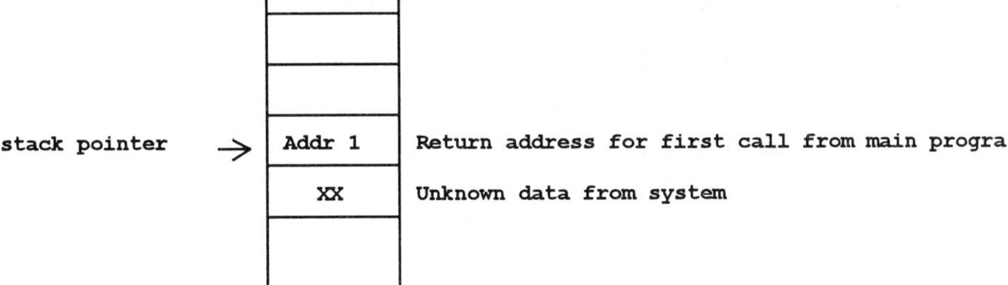

`3. After call to sub 2 from sub 1`

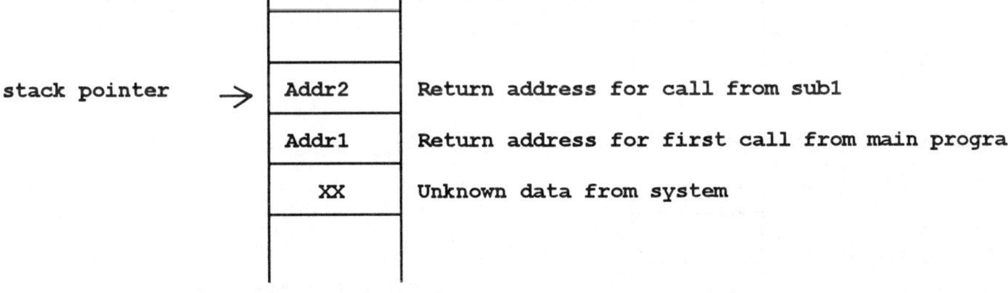

4. After return from sub 2

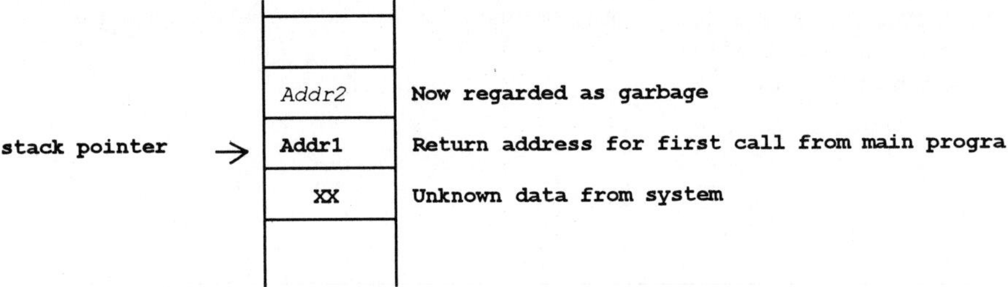

5. After first return from sub 1

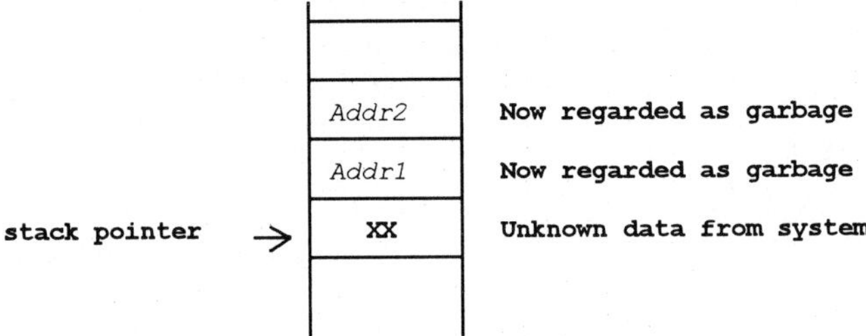

6. After second call to sub 1

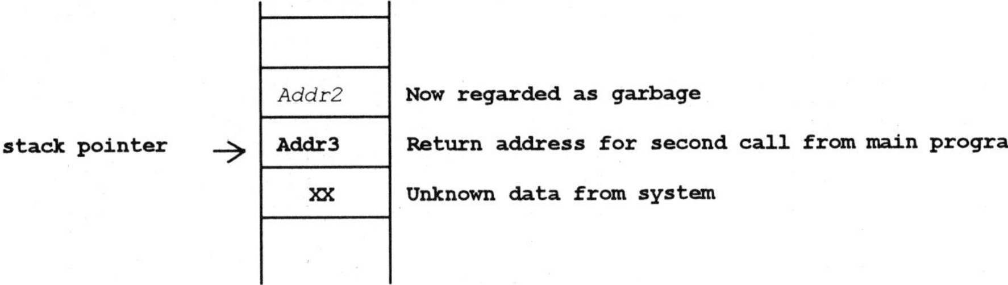

7. After call to sub 2 from sub 1

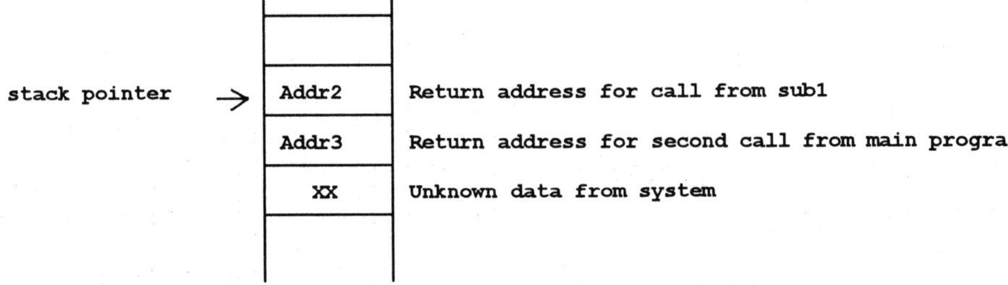

(The return from subroutine 2 back to subroutine 1 and then back to addr 3 in the main program follows the same process as steps 4 and 5.)

The instructions for calling a subroutine are:

```
        call    label                   (Z80 and 8086)
e.g.    call    sub1
```

```
        jsr     label                   (68000)
e.g.    jsr     sub1                     (JSR means Jump to SubRoutine)
```

The instructions for returning from a subroutine are:

```
        ret                             (Z80 and 8086)
                                        (RET means RETurn from subroutine)
```

```
        rts                             (68000)
                                        (RTS means ReTurn from  Subroutine)
```

9.4.2 Preserving Register Contents

Subroutines of necessity will alter the contents of one or more processor registers. When a subroutine is called from different parts of the main program there is a possibility that these registers will hold important data. To avoid loosing the data one of two approaches may be used:

either save the registers before calling the subroutine and retrieve them afterwards

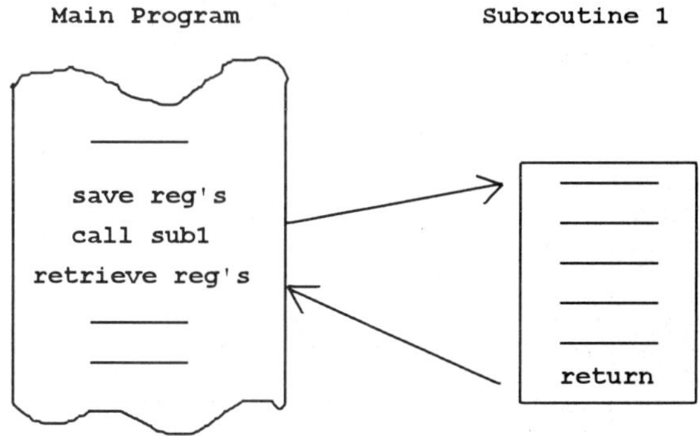

or save the registers on entering the subroutine and retrieve them just prior to leaving it.

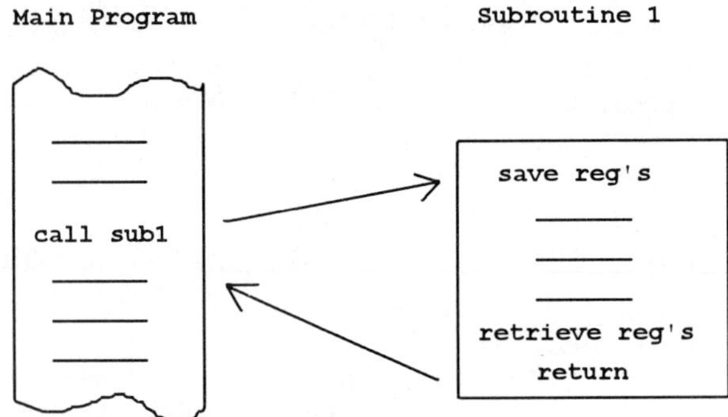

For this sort of purpose the most common place to save register contents is on the *stack*. The instructions for saving register contents on the stack are:

	push	*qq*	**(Z80)**
e.g.	**push**	**bc**	

	push	*reg*	**(8086)**
e.g.	**push**	**bx**	

	move	*Dn,-(A7)*	**(68000)**
e.g.	**move.w D2,-(A7)**		(You may also use *.b* and *.l*)

The instructions for retrieving register contents from the stack are:

	pop	*qq*	**(Z80)**
e.g.	**pop**	**bc**	

	pop	*reg*	**(8086)**
e.g.	**pop**	**bx**	

	move	*+(A7),Dn*	**(68000)**
e.g.	**move.w +(A7),D2**		(You may also use *.b* and *.l*)

Note that when saving registers on a Z80 it is always a 16 bit register pair such as bc or hl which is saved. On an 8086 it is a full 16-bit register such as ax or bx which is saved *(not* al or bh). On the 68000 however, the size of the data saved may be either byte or word or long word.

```
*************************************************************
*  VERY  IMPORTANT                                          *
*                                                           *
*  Remember that the stack is Last In, First Out. Thus, if the registers were  *
*  saved with:                                              *
*              push bc                                      *
*              push de                                      *
*                                                           *
*  they MUST be retrieved with                              *
*                                                           *
*              pop de         NOT         pop bc            *
*              pop bc                      pop de           *
*************************************************************
```

9.4.3 Parameter Passing

In many cases it will be necessary to pass information to a subroutine, a routine for calculating the square root of a number, for example, will need to know the number on which it is to work. Passing this sort of information to a subroutine is known as *parameter passing*. The two most often used techniques for parameter passing are:

    (a)   placing the data in one or more of the processors registers.
    (b)   placing the data in predetermined memory locations

A third method using the stack is also widely used but this is not recommended for the novice programmer.

Example 9.3
Show the call and return instructions that would be necessary to implement the program structure diagram for Example 5.3.2. The diagram is repeated below for convenience.

In the code which follows, the Z80 has been dealt with in detail but for the 8086 and the 68000 only the differences are indicated. You should also note that some assemblers place quite tight restrictions on the length of names and labels - sometimes as little as eleven characters. Thus, a label which might have been *p-time-act-body* to be consistent with the diagram, has been abbreviated to *p-tm-act-bd* to fit it into eleven characters.

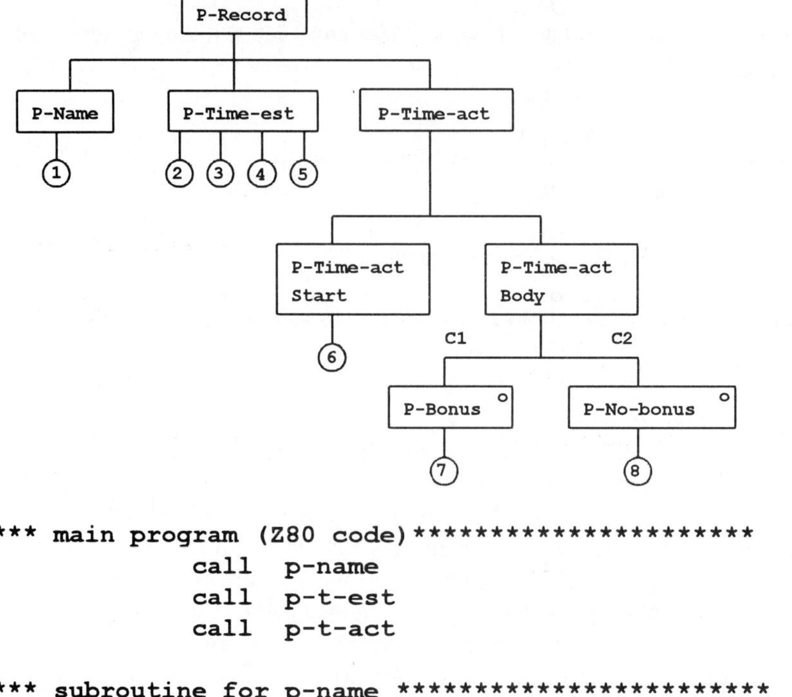


```
;******* main program (Z80 code)*********************
                call   p-name
                call   p-t-est
                call   p-t-act


;******* subroutine for p-name ***********************
p-name:                 .......
                                        ;code for operation 1

                        .......
                ret


;******* subroutine for p-tm-est *********************
p-tm-est:               .......
                                        ;code for operations 2,3,4,5

                        .......
                ret


;******* subroutine for p-tm-act *********************
p-tm-act:               call   p-tm-act-st
                        call   p-tm-act-bd
                ret


;******* subroutine for p-tm-act-st ******************
p-tm-act-st:            .......
                                        ;code for operation 6

                        .......
                ret
```

```
;******* subroutine for p-tm-act-bd ******************
p-tm-act-bd:            ld      a,bonus_time
                       cp      a,act_time
                       jp      c,no_extra
                       call    bonus
                       ret


no_extra:              call    no-bonus
                       ret


;******* subroutine for bonus ***********************
bonus:                 . . . . . . .
                                       ;code for operation 7
                       . . . . . . .
                       ret


;******* subroutine for no bonus ********************
no-bonus:              . . . . . . .
                                       ;code for operation 8
                       . . . . . . .
                       ret
```

For the 8086 processor the call and return from subroutines are identical. For the 68000 processor these are replaced with instructions such as:

```
               jsr    p-name
      and      rts
```

The 8086 version of **p-tm-act-bd** becomes:

```
;******* subroutine for p-tm-act-bd *****************
p-tm-act-bd:           mov     al,bonus_time
                       cmp     al,act_time
                       jc      no_extra
                       call    bonus
                       ret


no_extra:              call    no-bonus
                       ret
```

and the 68000 version becomes:

```
;******* subroutine for p-tm-act-bd ****************
p-tm-act-bd:        move.b bonus_time,D0
                    cmp    act_time,D0
                    bcs    no_extra
                    jsr    bonus
                    ret


no_extra:           jsr    no-bonus
                    ret
```

N.B. If we ignore what happens when **act_time = bonus_time** the subroutine **p-tm-act-bd** would have the same effect if it was written as:

```
;******* subroutine for p-tm-act-bd ****************
p-tm-act-bd:        ld     a,bonus_time
                    cp     a,act_time
                    jp     nc,extra
                    call   no-bonus
                    ret


extra:              call   bonus
                    ret
```

However, the left to right order on the program flow diagram puts *bonus* first and *no-bonus* second. Using the first version of *p-tm-act-bd* is consistent with this order. Maintaining such consistency can be of considerable help when trying to debug programs.

9.5 OTHER COMMON CONTROL STRUCTURES

Although not permitted by JSP there are two other control structures **do-while** and **for-next** which are in common use. These are shown in Figure 9.5.

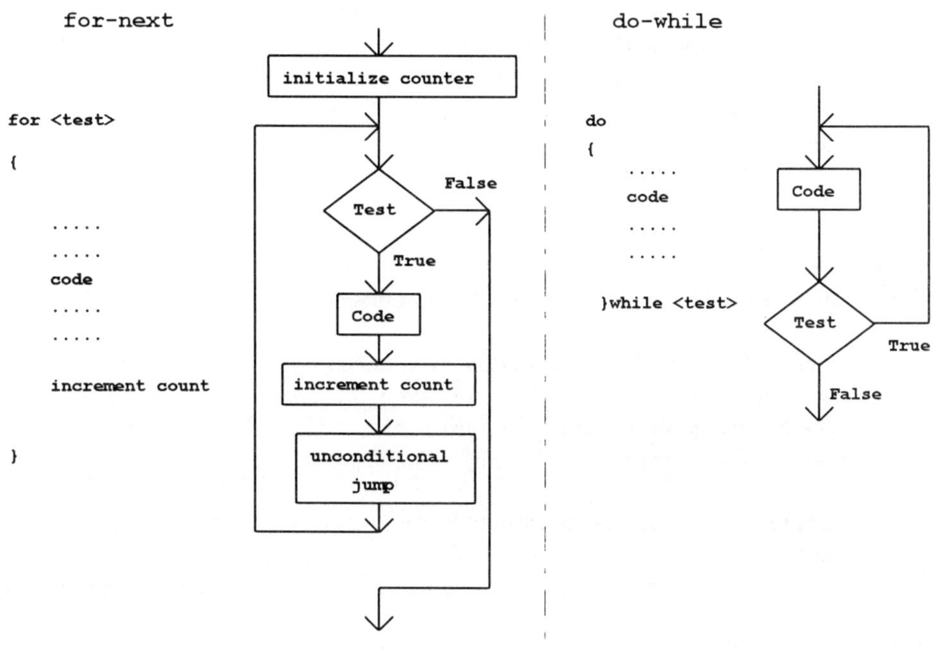

Figure 9.5

The for-next has a lot in common with the while loop. The main difference is that the for-next loop must initialize a counter at the start and must adjust the count before returning to the test.

The do-while loop is also similar to the while loop. The difference in this case is that do-while loop ensures that the code in the loop is executed at least once.

Exercise 9.1

An experimental weather station records air temperature at 30 minute intervals over a 6 hour period and logs the readings as 8 bit numbers in consecutive memory locations. The maximum temperature is then displayed. Write a program to simulate this by putting 12 bytes of trial data in consecutive memory locations and then search these location for the biggest number and display the result on the LEDs. The following pseudo code is offered as a starting point:

> set counter to zero
> point a register at the data items
> copy [pointer] into a register (call this ***temp_store***)
> loop while count not equal to 12
> > increment pointer
> > if [pointer] > [temp_store]
> > > copy [pointer] to [temp_store]
> > increment count
>
> end loop
> send [temp_store] to LEDs
> exit

Exercise 9.2 (for 80x86 users only)

A program is needed to meet the following requirements:

> *Using the 8086 skeleton program as a starting point adapt Example 6.2 so that, after the set up procedures, the main program calls a subroutine to read the number of cycles (**num_cyc**) from the switches. The main program must then call a second subroutine to generate the tone.*

(a)   Express the requirements in pseudo code using the control structures given above.

(b)   Translate the pseudo code into assembly language, test and debug the program.

N.B. The instructions `'sti'`, `'mov ax,4c00h'` and `'int 21h'` which appear at the end of Example 6.2 should form part of the main program not the subroutine.

Exercise 9.3 (Z80 and 68000 users)

A program is needed to meet the following requirements:

> *Using the Z80 or 68000 skeleton program as a starting point adapt Example 6.5 (Z80) or 6.8 (68000) so that, after the set up procedures, the main program calls a subroutine to read a value for the delay time (**del_time**) from the switches. The main program must then call a second subroutine to generate the sequence. After each sequence the program must read the switches again for a new delay time.*

(a)   Express the requirements in pseudo code using the control structures given above.

(b)   Translate the pseudo code into assembly language, test and debug the program.

10

Common Tasks and their Solution

10.1 ARITHMETIC MANIPULATIONS

10.1.1 Addition and Subtraction

Simple addition and subtraction has been covered in earlier chapters. Complications can arise if the result is outside the range that can be handled by the ALU. If errors are to be avoided it is essential that adequate preparation is made at the planning stage. Adding two 8-bit numbers could result in a 9-bit answer and storing this number would require two 8-bit memory locations.

Example 10.1.1

Write the pseudo code for a program which is to add together twenty 8-bit unsigned binary numbers stored in consecutive memory locations. The result is to be stored in consecutive memory locations. It must be possible to use the same pseudo code for either Z80 or 68000 or 8086.

Notes   1   If the code is to work on all three processors, we must stick to 8-bit arithmetic operations.

             2   The maximum value for any of the 8-bit numbers is ff_{16} or 255. Therefore the maximum total is $20 \times 255 = 5100$ $(= 13EC_{16})$ and this will require two memory locations which will be designated **res_lo** for the low byte of the result and **res_hi** for the high byte.

             3.   For maximum speed data should be in processor registers rather than memory. The original data is already in memory but we can keep the cumulative totals in registers. We will call the lower byte of the cumulative total **lct** and the high byte **hct**.

Thus the pseudo code could be:

```
clear a register to hold lct
clear a register to hold hct
point a register at the list of numbers
set counter to 20
```

```
while counter > 0
{
                lct = lct + [pointer]
                if carry set
                            increment hct
                increment pointer
                decrement counter
}
copy lct to res_lo
copy hct to res_hi
```

10.1.2 Multiplication

The 68000 and the 8086 have instructions to perform multiplication directly and if you want to use these you should consult the instruction set. The Z80 has no such instructions and it is necessary to create the multiply function using a technique such as the repeated add for which the pseudo code was given in Chapter 5. When there is a need to multiply by a power of 2 (2, 4, 8, 16 etc.) it should be noted that shifting the data one place left and shifting a zero into the least significant bit is the same as multiplying by 2.

e.g. $0000\ 0101_2$ $(=5_{10})$ shifted left becomes $0000\ 1010_2$ $(=10_{10})$

and this is usually the fastest way of multiplying. Whatever procedure is being used it is important to ensure that adequate space is allocated for the maximum possible result.

10.1.3 Division

The 68000 and the 8086 have instructions to perform division directly and if you want to use these you should consult the instruction set. The Z80 has no such instructions and it is necessary to create the division function using a technique such as repeated subtraction or, if the divisor is a power of two, then shift right may be used. Whatever technique is used there will be a need to store both the quotient and the remainder. The basic process for division by repeated subtraction is:

```
clear a register for quotient  (Q)
copy dividend into a register (DIVD)
copy divisor into a register (DIVS)
clear carry flag
while DIVD >= 0
{
        Q = Q + 1
        DIVD = DIVD - DIVS
}
DIVD = DIVD + DIVS
Q = Q - 1
```

i.e.   Count the number of times you can take DIVS from DIVD without getting a negative result. Once a negative result has occurred you have performed one subtraction too many so add one DIVS back to DIVD and decrement the count. Q then holds the quotient and DIVD holds the remainder.

10.1.4 Average

In control applications the calculation of an average is quite common. In automatic packaging, for example, the readings from a transducer indicating the weight of substance will be affected by mechanical vibrations. To reduce the risk of error a number of readings (samples) are taken and the average calculated. If you are designing such a system it is worth considering making the number of samples a power of two. This will enable division to be performed using shift operations instead of the slower repeated subtraction operation.

Now complete Exercise 10.1 at the end of this chapter

10.2 TIME DELAYS

Time delays form an important part of many programs for microprocessor based systems. They are necessary for one of two main reasons:

   (a)   the equipment being controlled is not capable of operating at the same sort of speed as the microprocessor and we must force the microprocessor to waste time. This is often the case when the microprocessor is controlling mechanical or electromechanical devices.
   (b)   the process being controlled requires a specific time delay e.g. turn a heater on for five minutes and turn it off again.

Whatever the reason there are two main ways of creating a time delay:

   (a)   make the processor count down to zero from a given value.
   (b)   use some sort of timer I.C. of which there are many which might be used.

In this chapter we shall deal only with (a). The basic form of a time delay loop is:

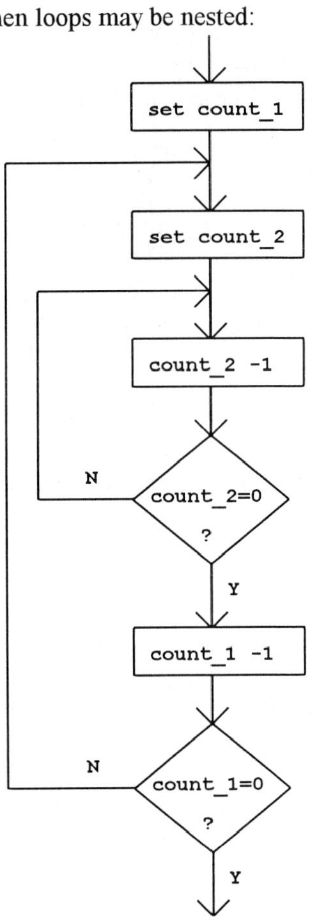


```
set count value

do
{
        count = count - 1


} while count not = 0
```

If this does not produce a delay of sufficient length then loops may be nested:

```
set count_1 value

do

{
        set count_2 value


        do
        {
                count_2 = count_2 - 1


        } while count_2 not = 0


        count_1 = count_1 - 1


} while count_1 not = 0
```

Example 10.2.1
Refer back to Example 6.4 (80x86) or Example 6.7 (Z80) or Example 6.10 (68000)
and identify the time delay. Show how the total time for each of these may be increased
by a factor of 5.

In each of the *proposed* solutions below the original time delay is shown in *light italics*
and the new loop is shown in **bold** typeface

8086

```
                mov     dl,5h
loop4:          mov     al,10h
loop3:          mov     bl,0h
loop2:          mov     cl,0h
loop1:          dec     cl
                jne     loop1
                dec     bl
                jne     loop2
                dec     al
                jne     loop3
                dec     dl
                jne     loop4
                ret
```

Z80

```
                ld      c,5
loop3:          ld      b,del_time
loop2:          ld      a,0
loop1:          dec     a
                jp      nz,loop1
                dec     b
                jp      nz,loop2
                dec     c
                jp      nz,loop3
                ret
```

68000

```
                move.b  #5,d1
loop2           move.l  #del_time,d0
loop1           sub.l   #1,d0
                bne.s   loop1
                sub.b   #1,d1
                bne.s   loop2
                rts
```

These proposed solutions might appear satisfactory but it should be noted that, although the main body of the delay routine is executed five times, no allowance has been made for the extra instruction being used. The total delay time will therefore be slightly more than five time the original delay time. For many applications this error will be totally insignificant - in a washing machine program for example it doesn't really matter if the washing spins for 5 minutes or 5 minutes 10 seconds. For other applications far greater accuracy is required and it will be necessary to determine the number of clock cycles taken for the entire routine. If we know the number of clock cycles and the frequency of the system clock we can calculate the total time from:

$$\text{total time} = \text{number of clock cycles} \times \frac{1}{\text{clock frequency}}.$$

If you examine the instruction set you will see that it quotes the number of cycles taken for each instruction:

		No. clock cycles
Z80	`ld  r,n`	7
	`dec`	4
	`jp cc,nn`	10

(Clock cycles are referred to as T states in the Z80 instruction set)

68000	`move.b immed,Dn`	8
	`move.w immed,Dn`	8
	`move.l immed,Dn`	12
	`sub.b  immed,Dn`	8
	`sub.w  immed,Dn`	8
	`sub.l  immed,Dn`	16
	`bcc    nn`	10 if condition met
		8 if condition not met

8086	`mov  reg,immed`	4 (2 on 286 and 386)
	`dec  reg8`	3 (2 on 286 and 386)
	`jcc  label`	16 if condition met (7 on 286 and 386)
		4 if condition not met (3 on 286 and 386)

Example 10.2.2

Calculate the execution times for each of the following routines:

(a)
```
                  ld    a,80h
     loop1        dec   a
                  jp    nz,loop1
```

for a Z80 with a 1.79 MHz clock.

(b)
```
                    move.b   #$80,D0
        loop1       sub.b    #$1,D0
                    bne.s    loop1
```

for a 68000 with an 8 MHz clock.

(c)
```
                    mov      al,80h
        loop1:      dec      al
                    jne      loop1
```

for an 80286 with a 16 MHz clock.

In all three cases the conditions for the jump will be met on 127 (7F hex) occasions and will not be met on one occasion. Note that with both the 80286 and the 68000, there is a difference in the number of clock cycles required when the jump is taken and when it is not taken. Thus the total times are:

Z80

$$(7 + (4 + 10)128 \ \times \ \frac{1}{1.79 \times 10^6} \ = 1.005 \times 10^{-3} \text{ s} \ = 1.005 \text{ ms.}$$

68000

$$(8 + (8 + 10)127 + (8 + 8)) \ \times \ \frac{1}{8 \times 10^6} \ = \ 0.289 \times 10^{-3} \text{ s} = 0.289 \text{ ms.}$$

80286

$$(2 + (2 + 7)127 + (2 + 3)) \ \times \ \frac{1}{16 \times 10^6} = 71.9 \times 10^{-6} \text{ s} = 71.9 \text{ } \mu\text{s}$$

Note for 8086 users
There is an instruction **loop** which combines a **16 bit dec** with a **jne**. It has the format:

	No clock cycles	
loop *label*	8086	17 if taken
		5 if not taken
	80286	8 + m if taken
		4 if not taken

(m = number of bytes in next instruction)

The contents of register CX are decremented and, if this does not produce a zero result, the loop or jump to the label is taken.

Example 10.2.2
Use the results of Example 10.2.1 to make an order of magnitude assessment of the length of time to execute one instruction for each of the three systems.

In all three programs :

loading a register	is done once
decrementing a register	is done 128 (80 hex) times
jump or branch	is done 128 times (127 times the branch is taken and on one occasion it is not taken)

$\therefore$	total number of instructions executed $= 1 + 128 + 128 = 257$

Z80 approx. time for 1 instruction $= \dfrac{1.005 \times 10^{-3} \text{ s}}{257} = 3.9 \times 10^{-6} \text{ s} = \textbf{3.9 } \boldsymbol{\mu}\textbf{s}$

68000 approx. time for 1 instruction $= \dfrac{0.289 \times 10^{-3} \text{ s}}{257} = 1.1 \times 10^{-6} \text{ s} = \textbf{1.1 } \boldsymbol{\mu}\textbf{s}$

80286 approx. time for 1 instruction $= \dfrac{71.9 \times 10^{-6} \text{ s}}{257} = 0.3 \times 10^{-6} \text{ s} = \textbf{0.3 } \boldsymbol{\mu}\textbf{s}$

N.B. In Example 10.2.3 below you are recommended to read through the solutions for all the processors since they each show a slightly different aspect of calculations for time delay.

Example 10.2.3
A games show gives contestants 10 seconds in which to answer a question. Write a program which will start timing as soon as the question master presses a button and which will indicate *time out* by turning on an LED. Use the system data from Example 10.2.1. and assume that an accuracy of +/- 0.1 s is required.

The basic program requirements are:

```
while switch not closed
{
        read switch
}
delay 10 seconds
illuminate LED
```

From Example 10.2.2 it can be seen that, even for the slowest processor, the time taken to:

(a)   leave the *read switch* routine and enter the delay,

and   (b)   leave the delay and illuminate the LED,

will be negligible (a few tens of microseconds each) compared to the accuracy required for the program .   We can therefore regard the 10 second time interval as being dependent only on the time in the delay loop. Our first task is to estimate the number of loops required and then to determine more accurately the values in the loop counters.

Z80

(a)   Z80 registers are 8 bits wide and so can have a maximum value of 255. If we arrange the delay loops so that the outer loop is executed 200 times then we should be able to achieve an accuracy of 10 s/ 200 = 0.05 s which is well within the specification. This implies that the inner loop or loops must take 0.05s.

(b)   From the work in Example 10.2.1 the maximum time we can get from a single loop is 1.005 ms x 2 = 2.01 ms. To create a delay of 0.05 s would require the innermost loop to be executed $0.05/ 2.01 \times 10^{-3} = 24.8$ say, 25 times.

(c)   The previous two paragraphs give us our first estimates for the loops:

(i)   inner loop (loop1) - set loop counter to 0 so that the loop is executed 256 times (first decrement instruction will take the register to ff hex).

(ii)   middle loop (loop2) - set loop counter to 19 hex (25 decimal).

(iii)   outer loop (loop3) -set loop counter to C8 hex (200 decimal).

(d)   We now calculate the delay produced by these estimates:

```
                                        No. T states
                ld      a,0c8h          ; 7
        loop3:  ld      b,19h           ; 7
        loop2:  ld      c,0             ; 7
        loop1:  dec     c               ; 4
                jp      nz,loop1        ; 10
                dec     b               ; 4
                jp      nz,loop2        ; 10
                dec     a               ; 4
                jp      nz,loop3        ; 10
```

loop1   This consists of:

            dec c   jp nz,loop1
            (4)   (10)

Register C is loaded with 0 and so the maximum count will be 256. Of these the jump will be taken 255 times and on the last time through the loop, the jump will not be taken.

$\therefore$ No. cycles in loop 1 = (4 + 10)256 = 3584

loop 2   This consists of:

ld c,0	loop1	dec b	jp, nz,loop2
(7)	(3584)	(4)	(10)

Register B is loaded with 19 hex (25 decimal)

$\therefore$ No. cycles in loop2 = (7 + 3584 + 4 + 10)25 = 90125

loop3   This consists of

ld b,16h	loop2	dec a	jp nz,loop
(7)	(90125)	(4)	(10)

Register A is loaded with c8 hex (200 decimal)

$\therefore$ No. cycles in loop3 = (7 + 90125 + 4 + 10)200 = 18029200

Total   Total delay consists of :

ld a,0c8h	loop1
(7)	(18029200)

$\therefore$ No. cycles in delay = 7 + 18029200 = 18029207

and at 1.79 MHz total delay time = 18029207 x $(1.79 \times 10^6)^{-1}$ = 10.07 s

This is well within specification

6800

A simple loop will consist of two instructions (decrement and jump) and so from Example 10.2.2 a simple loop should take about 2.2 μs. For a 10 second delay we would need to execute this loop 10/ 2.2 $\times 10^{-6}$ $\simeq$ 4.5 $\times 10^6$ times. The 68000 has 32 bit internal registers which means that the maximum count value in a register will be 2^{32} $\simeq$ 4.3 x 10^9 and so a single loop will suffice. The time taken to load the register used for a loop counter will be negligible compared to the total delay time. Also, since the loop is not nested, the difference between the number of cycles when the jump is taken and when it is not taken will be negligible. Thus our first estimate becomes:

No. cycles in one loop   = cycles for move.1 + cycles for sub.1

$\qquad\qquad$ =   12   +   16   = 28

$\therefore$ time for one loop   = 28 x $(8 \times 10^6)^{-1}$ = 3.5 x 10^{-6}

$\therefore$ No. times loop needs to be executed = 10 /3.5 x 10^{-6} = 2857142.9

$\qquad\qquad\qquad\qquad\qquad\qquad\qquad$ say   2857143

$\qquad\qquad\qquad\qquad\qquad\qquad\qquad\qquad\qquad$ (2b98b7 hex)

Hence total time in loop = 28 x 2857143 x $(8 \times 10^6)^{-1}$ = 10.00 s (to 2 dec places)

80286

From the work in Example 10.2.2 the two instructions in a simple loop should take about 0.6 µs. Thus we need to execute the loop $10/0.6$ µs $\simeq 17 \times 10^7$ times. The 80286 has 16 bit registers which means that the maximum count value will be 65,536. Nesting two loops would enable us to get an effective count of $2^{32} \simeq 4.3 \times 10^9$ and so two loops will suffice. If we arrange the delay loops so that the outer loop is executed 50,000 times then we should be able to achieve an accuracy of 10 s$/50,000 = 0.0002$ s which is well within the specification. If the basic program is :

```
                                           No. cycles
              mov     ax,num1       ;  2
loop2:        mov     bx,num2       ;  2
loop1:        dec     bx            ;  2
              jne     loop1         ;  7/3
              dec     ax            ;  2
              jne     loop2         ;  7/3
```

then:

$$\text{loop1 takes } (2 + 7) / 16 \times 10^6 \qquad = 0.5625 \text{ µs}$$
$$\therefore \qquad \text{num1} \qquad = 0.0002/0.5625 \text{ µs}$$
$$= 355.6 \text{ say } 356$$
$$(164 \text{ hex})$$

Using this estimate we have:

No. cycles in loop1 $= (2 + 7)355 + 2 + 3 = 3200$
No. cycles in loop2 $= (2 + 3200 + 2 + 7)49,999 + 2 + 3200 + 2 + 3$
$\qquad\qquad\qquad = 160549996$
No. cycles in delay $= 160549996 + 2 = 160549998$

and at 16 MHz total delay time $= 160549998 \times (16 \times 10^6)^{-1} = 10.03$ s

Although we could clearly get this much closer, we are well within specification and there is little point in wasting time on further calculations.

Now complete Exercises 10.2 and 10.3 at the end of this chapter

10.3 PULSE COUNTING

The need to count pulses occurs in many situations e.g.

(a)   counting cars entering and leaving a car park so that a "CAR PARK FULL" sign may be illuminated when necessary.

(b)   counting items passing along a conveyor belt so that they may be boxed in appropriate quantities.

    (c)   counting the output pulses from a Geiger-Muller tube so that the level of radioactivity may be indicated.

In microprocessor based systems pulse counting is usually achieved with *either* a standard I/O device and appropriate software *or* using hardware involving some sort of a counter I.C. In this chapter we shall deal only with the former. Whatever the source of the signal, basic pulse counting consists of:

    (a)   waiting for the pulse to go high.
    (b)   waiting for the pulse to go low
    (c)   incrementing the count.

In pseudo code this becomes:

```
while pulse low
{
        read pulse input
}
while pulse high
{
        read pulse input
}
increment count.
```

Provided that the pulse is 'clean', i.e. it has clear well defined edges as shown in Figure 10.1.(a), then this basic approach is quite satisfactory. However, the contacts of mechanical switches bounce and so opening and closing a switch will give rise to the sort of pulse shown in Figure 10.1.(b) (known as **switch bounce** or **contact bounce**). Also, rapidly changing electronic signals combined with capacitance and inductance in a circuit can give rise to the phenomenon known as **ringing** which is illustrated in Figure 10.1.(c).

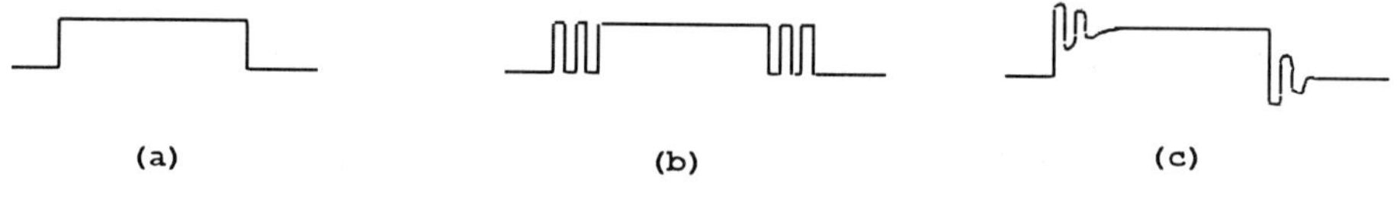

Figure 10.1

Both switch bounce and ringing may be either eliminated or reduced to an acceptable level with hardware but, for many applications, this just adds unnecessary cost. A cheap and effective software solution involves the use of a time delay to allow the change in signal level time to stabilize before moving to the next stage of the process. In pseudo code the basic pulse counting routine becomes:

```
while pulse low
{
        read pulse input
}
delay
while pulse high
{
        read pulse input
}
delay
increment count.
```

Example 10.3.1

Write a program to count the number of times that the switch at bit 7 is operated and display the count on the LEDs.

The pseudo code might be:

```
set count to zero
clear display
do
{
        do
        {       read switches
                mask for bit 7
        }while switch open
        call delay
        do
        {       read switches
                mask for bit 7
        } while switch closed
        call delay
        increment count
        display count
}
```

Hence the additions to the skeleton files might be:

Z80

```
********** add user code between these lines ***********
                ld      b,0             ;clear count
                ld      a,0             ;clear display
                out     (led),a         ;           "
read1           in      a,(sw)          ;read switches
                and     80h             ;mask for bit 7
                jp      z,read1         ;jump if open
                call    delay
read2           in      a,(sw)          ;read switches
                and     80h             ;maks for bit 7
                jp      nz,read2        ;jump if closed
                call    delay
                inc     b               ;increment count
                ld      a,b             ;copy count to acc
                out     (led),a         ;send to LEDs
                jp      read1

delay           ;** see note below **
                ret

;*****************************************************************
exit            rst     30h
```

68000

```
***********     user code between these lines **************
                move.b  #$0,d0          *clear count
                move.b  #$0,led         *clear display
read1           move.b  sw,d1           *read switches
                and.b   #$80,d1         *mask for bit 7
                beq.s   read1           *jump if open
                jsr     delay           *call delay
read2           move.b  sw,d1           *read switches
                and.b   #$80,d1         *mask for bit 7
                bne.s   read2           *jump if closed
                jsr     delay           *call delay
                add     #$1,d0          *increment count
                move.b  d0,led          *display count
                jmp     read1

delay       ;** see note below **
                rts
***************************************************************
exit            trap    #11 .
```

8086

```
;** enter user data between these lines **
        mov     cl,0            ;count = 0
        mov     al,0            ;clear display
        mov     dx,[o_port]     ;       "
        out     dx,al           ;       "

read1:  mov     dx,[i_port]     ;read switches
        in      al,dx           ;       "
        and     al,80h          ;mask for bit 7
        je      read1           ;jump if bit 7 = 0
        call    delay

read2:  mov     dx,[i_port]     ;read switches
        in      al,dx           ;       "
        and     al,80h          ;mask for bit 7
        jne     read2           ;jump if bit 7 not = 0
        call    delay

        inc     cx              ;increment count
        mov     ax,cx           ;copy count to ax
        mov     dx,[o_port]     ;display count
        out     dx,al           ;       "
        jmp     read1

delay:          ;** see note below **
        ret
;*****************************************
```

Note. It is left to the reader to convert one of the time delay routines given earlier in this chapter into a suitable subroutine. The length of delay required can only be determined by experiment. However, experience shows that, with switches designed for light current work, a delay of a couple milliseconds is more than adequate.

Now complete Exercise 10.4 at the end of this chapter

10.4 WAVEFORM GENERATION

With suitable additional hardware, microprocessor based systems may be used to produce waveforms of almost any shape. Without such hardware it is only really the mark-to-space ratio of a rectangular wave which may be altered. Altering the mark-to-space ratio is quite often used for things such as light dimmers, speed control of small dc motors etc..

Example 10.3
Write a program which will illuminate two adjacent LEDs. The first LED is to be illuminated at full brightness, the second is to be illuminated at roughly quarter brightness.

For roughly quarter brightness we will need roughly a quarter of the energy. One way of achieving this is to turn the LED on for a quarter of the time and off for three quarters of the time. If this is done sufficiently quickly it will appear that the LED is continuously illuminated but at a reduced brightness. If we use bits 6 and 7 the pseudo code might be:

```
        illuminate both LEDs
        do
        {
                delay
                change the state of LED 7
                delay for 3 times delay 1
                change state of LED 67
        }
```

We have chosen to "change the state of LED 7" rather than specifying a particular logic level so as to illustrate the use of XOR for this purpose. The corresponding additions to the skeleton file might be:

Z80

```
;********** add user code between these lines ***********
            ld      a,0c0h      ;turn both LEDs ON
            out     (led),a     ;           "

repeat      ld      c,20h       ;loop counter for time ON
            call    delay
            xor     80h         ;invert bit 7 in acc
            out     (led),a     ;send to LEDs
            ld      c,60h       ;loop counter for time OFF
            call    delay
            xor     80h         ;invert bit 7 in acc
            out     (led),a     ;send to LEDs
            jp      repeat

delay       dec     c
            jp      nz,delay
            ret
;***************************************************************
exit        rst     30h
```

68000

```
***********  user code between these lines  ***************
            move.b   #$c0,led   *turn both LEDs On
            move.b   #$80,d0    *pattern for inverting in D0

repeat      move.l   #$200,d2   *loop counter for time ON
            jsr      delay
            eor.b    d0,led     *invert LED 7
            move.w   #$600,d2   *loop counter for time OFF
            jsr      delay
            eor.b    d0,led     *invert LED 7
            jmp      repeat

delay       sub.w    #$1,d2
            bne.s    delay
            rts
***********************************************************************
```

8086

```
;******** enter user code between these lines  *************
            mov      al,0c0h          ;bit pattern for both ON
            mov      dx,[o_port]      ;send to LEDs
            out      dx,al            ;       "

repeat:     mov      cx,200h          ;loop counter for time ON
            call     delay
            xor      al,80h           ;invert bit 7
            mov      dx,[o_port]      ;send to LEDs
            out      dx,al            ;       "
            mov      cx,600h          ;loop counter for time OFF
            call     delay
            xor      al,80h           ;invert bit 7
            mov      dx,[o_port]      ;send to LEDs
            out      dx,al            ;       "
            jmp      repeat

delay:      loop     delay
            ret
;*******************************************
```

(It is suggested that you experiment with the values of the loop counters so as to vary
both the total cycle time and the mark-to-space ratio. As you vary the total cycle time
try to establish the point at which you are able to detect flicker on the LED which is
being turned on and off.)

Now complete Exercise 10.5 at the end of this chapter

10.5 SERIAL INPUT AND OUTPUT

As you have already seen data is passed around a typical microprocessor based system eight (or sixteen) bits at a time. If the system is to communicate with a peripheral or another microprocessor based system which is in close proximity then sending all eight bits at one time is probably the best approach. This is known as **parallel** communication. If however the peripheral is some distance away then cabling costs might make it more economical to send the data one bit at a time down a single data line. This is known as **serial** communication. The main penalty of serial communication is that hardware and/or software is needed at the transmitting end to break the parallel data down into serial data and at the receiving end to re compose serial data into parallel data i.e.

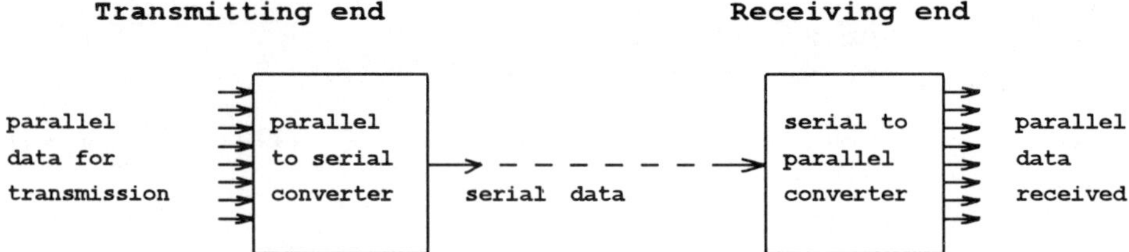

If we wanted to send the data *least significant bit first* we can make use of the processor's rotate right instructions (RRA - Z80, ROXR - 68000, RCR 8086). The effect of these instructions are:

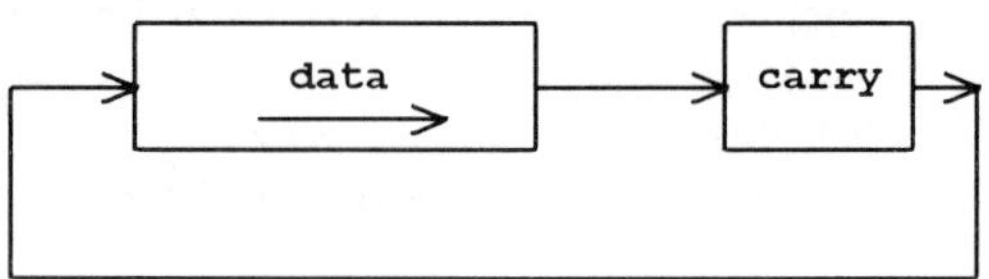

Thus when transmitting data we have:

	data	**carry**	
original data	01010101	x	(not transmitted)
shift once	x0101010	1	(to be transmitted)
shift again	xx010101	0	(to be transmitted)
shift again	xxx01010	1	(to be transmitted)
shift once	xxxx0101	0	(to be transmitted)
shift again	xxxxx010	1	(to be transmitted)
shift again	xxxxxx01	0	(to be transmitted)
shift again	xxxxxxx0	1	(to be transmitted)
shift again	xxxxxxxx	0	(to be transmitted)

The data being shifted back in to the most significant end has been shown as "x" since, although it is predictable, it is not relevant.

When transmitting serial data there will usually be a need to include a time delay. Thus, to transmit one byte of data, the pseudo code might be:

```
set counter to 8
while counter not zero
{
        shift data right
        if carry set
                transmit logical 1
        else
                transmit logical 0
        call delay
        decrement count
}
```

Receiving serial data and converting back to parallel data may be achieved in a similar way. This time we use instructions to set or clear the carry flag according to the data received. The pseudo code might be:

```
set counter to 8
while counter not zero
{
        read serial data bit
        if serial data logical 1
                shift in a logical 1
        else
                shift in a logical 0
        call delay
        decrement count
}
```

If the serial data received is 55 hex (0101 0101 binary) we would have:

input	data in register	
1	xxxxxxxx	shift once
0	1xxxxxxx	shift again
1	01xxxxxx	shift again
0	101xxxxx	shift again
1	0101xxxx	shift again
0	10101xxx	shift again
1	010101xx	shift again
0	1010101x	shift again
x	01010101	

Example 10.4

Write a program which will treat bit 7 of the switches as a serial input, read this input eight times at 2 second intervals, convert it to parallel data and display the accumulating parallel data on the LEDs.

The pseudo code might be:

```
        clear a register for parallel data
        clear display
        set counter to 8
        do
        {       read switches
                mask for bit 7
                if serial data logical 1
                        shift in a 1
                else
                        shift in a 0
                display on LEDs
                call delay
                decrement count
        } while counter not zero
```

Z80

```
;********** add user code between these lines ***********
                ld      b,0             ;clear for para. data
                ld      a,0             ;clear display
                out     (led),a         ;         "
                ld      c,8             ;set count to 8
next            in      a,(sw)          ;read switches
                and     80h             ;mask for bit 7
                call    nz,shift_in_1
                call    z,shift_in_0
                ld      a,b             ; copy para. data to
acc
                out     (led),a         ;send to LEDs
                call    delay
                dec     c               ;decrement count
                jp      nz,next
                jp      exit


shift_in_1      scf                     ;set carry flag
                rr      b               ;shift into reg b
                ret
```

```
shift_in_0         scf                       ;set carry flag
                   ccf                       ;invert carry flag
                   rr       b                ;shift into reg b
                   ret

delay              ld       d,4
loop1              ld       e,0
loop2              ld       h,0
loop3              dec      h
                   jp       nz,loop3
                   dec      e
                   jp       nz,loop2
                   dec      d
                   jp       nz,loop1
                   ret
;**********************************************************************
exit               rst      30h


68000
***********    user code between these lines  ***************
                   move.b   #0,d0            *parallel data
                   move.b   #0,led           *clear display

                   move.b   #8,d1            *set count to 8
next               move.b   sw,d2            *read switches
                   and.b    #$80,d2          *mask for bit 7
                   beq.s    logical_0        *check if data 0
                   jsr      shift_in_1       *jump if data 1
                   bra.s    display          *
logical_0          jsr      shift_in_0       *jump if data 0
display            move.b   d0,led           *display on LEDs
                   jsr      delay
                   sub.b    #1,d1            *decrement count
                   bne.s     next
                   bra.s    exit

shift_in_1         ori.b    #$10,SR          *set X flag in sr
                   roxr.b   #1,d0            *shift data into d0
                   rts

shift_in_0         andi.b   #$ef,SR          *clear X flag in sr
                   roxr.b   #1,d0            *shift dat into d0
                   rts
```

```
delay           move.l   #$8b824,d7
loop            sub.l    #1,d7
                bne.s    loop
                rts


*************************************************************
exit            trap     #11
```


8086

```
;** enter user data between these lines **
                mov    cl,0          ;clear reg. for parallel data
                mov    al,0          ;clear display
                mov    dx,[o_port]   ;         "
                out    dx,al         ;
                mov    bx,8          ;set count = 8

next:           mov    dx,[i_port]   ;read switches
                in     al,dx         ;         "
                and    al,80h        ;mask for bit 7
                je     logical_0
                call   shift_in_1
                jmp    display
logical_0:      call   shift_in_0

display:        mov    al,cl         ;copy parallel data to al
                mov    dx,[o_port]   ;send to LEDs
                out    dx,al         ;         "
                call   delay
                dec    bl
                jne    next
                jmp    exit

shift_in_1: stc                      ;set carry flag
                rcr    cl,1          ;shift into reg cl
                ret

shift_in_0: clc                      ;clear carry flag
                rcr    cl,1          ;shift into reg cl
                ret
```

```
delay:          push  bx                    ;authors routine altered bx

        ;insert your 2 second delay routine here

                pop   bx
                ret
;* * * * * * * * * * * * * * * * * * * * * * * * * * * * * * * * * * *
```

Exercise 10.1
Write a program to determine the average of eight unsigned 8-bit numbers stored in consecutive memory locations and display the result on the LEDs.

Exercise 10.2
Refer to Case Study 1. Write a program to implement routine (b). The program should be written as a stand alone item and you should ignore the initial need for the switches to be at 01.

Exercise 10.3   *** *suggested for portfolio* ***
Refer to Case Study 2. Extend your work on this case study to include the entry and exit point. Remember that Circuit 1 should have no effect for the first 20 seconds after the system is armed and, if the circuit is broken after this time, a further 20 seconds should be allowed for the system to be disabled.

Exercise 10.4
Refer to Case Study 1. Write a program to implement routine (c). The program should be written as a stand alone item and you should ignore the initial need for the switches to be at 10.

Exercise 10.5   *** *suggested for portfolio* ***
A zoo wishes to make its normally nocternal animals active during day light hours so that they may be seen by visitors. To achieve this the animals have been placed underground and a microprocessor based system is used to control the lighting levels as follows:

> Normal nightime - lights on full to simulate daytime.
> Normal dawn - lights gradually dimmed to simulate night fall.
> Normal daytime - lights dim to simulate moon light.
> Normal dusk - lights gradually brightened to simulate sunrise.

Write a program to simulate the changes in light level at dawn and dusk. The final program is to take an hour to change between simulated daylight and simulated moon light but you should establish the principle by making the changes occur over about 10 seconds.

11

Interfacing

In the context of microprocessor based systems the term **interfacing** refers to the physical connections between the systems involved and the way in which data is to be exchanged. For many applications, the problems of interfacing have already been covered by earlier chapters. However, in the earlier exercises, you were able to rely on:

    (a)   the processor being available to dealt exclusively with the problem on which you were working.

    (b)   the devices with which the processor was communicating were either simple LEDs and switches, or could be treated as such, and therefore had no timing problems or communication protocols which needed to be established.

    (c)   all signals involved in data exchange were electronically totally compatible.

This chapter addresses some of the problems which occur when one or more of these conditions is not met.

Throughout most of this chapter we will use the terms **transmitter** and **receiver**. When a microprocessor based system is sending data to an external device, the microprocessor based system is the transmitter and the external device is the receiver. When the microprocessor based system is receiving data from an external device it is the microprocessor based system which is the receiver and the external device which is the transmitter.

11.1 SERIAL DATA TRANSFER TIMING

In the last chapter Example 10.4 illustrated one of the main problems associated with serial data transfer namely timing. There are three aspects to timing:

    (1)   knowing when the data is starting to arrive.

    (2)   being sure that the receiver is ready to receive the data.

    (3)   once data has started to arrive, knowing where one bit finishes and the next bit starts.

The first of these problems is usually solved by:

    either   using some feature of the incoming signal to indicate that new data is arriving (see section 11.1.2).

    or   using an additional control line to send a signal to the receiver that data is about to arrive (see section 11.2.2).

For many simple applications it is possible to ensure that the receiver is ready to receive data and so the second of these problems (2) may be ignored. This is illustrated by

Example 10.4 where the LEDs (the receiver) were always available to accept incoming data. When the problem cannot be ignored, the most common solution is to use another control line so that the receiver can signal to the transmitter and indicate when it is ready to receive data (see section 11.2.2).

The third of these problems (3) is associated with the speed at which the data is to be sent. The problem is usually overcome by using a predetermined data transfer rate or **baud rate** where:

$$\text{baud rate} = \frac{1}{\text{time for 1 bit}}.$$

and making sure that the receiver is set up to receive data at the specified baud rate. On occasions however the receiver may not have an internal clock and it is necessary to send a timing signal in addition to the data and any other control signals.

11.1.2 Asynchronous Serial Communication

The simplest serial communication uses just two wires - a signal wire and a ground wire i.e.

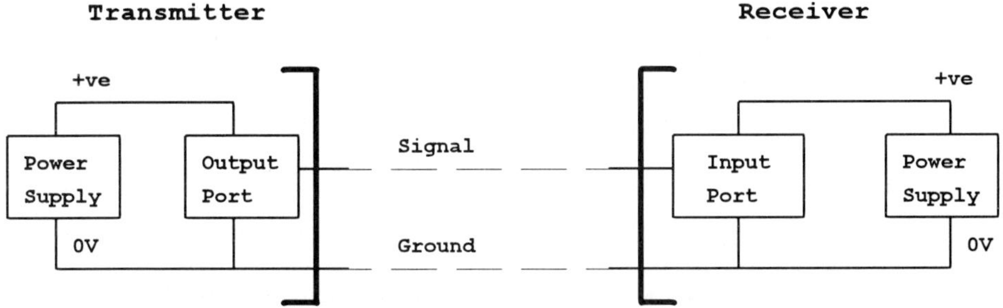

Figure 11.1

For this to be effective both systems must be set up for a common data transmission rate and a common data format. The most common format consists of a **start bit**, either 7 or 8 **data bits,** possibly a **parity bit** and either 1 or 2 **stop bits**. Start bits are normally a logical zero, stop bits are normally logical 1. If, for example, we wished to transmit data at 2400 baud and the format was to be 1 start bit, 8 data bits, no parity bit and 2 stop bits, the format for data transmission would be:

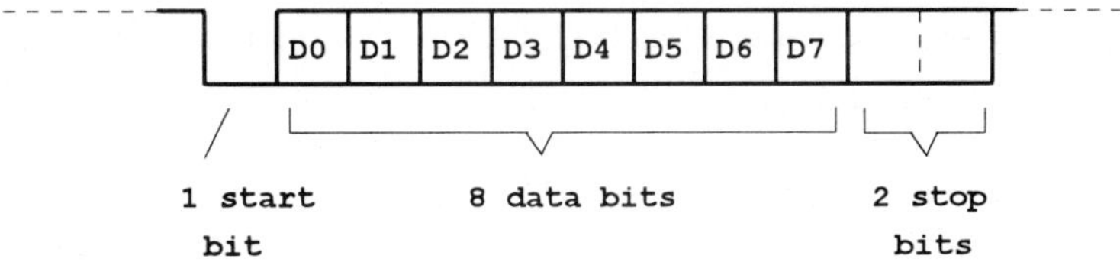

Figure 11.2

The receiver must be capable of reading its input port at speeds very much greater than the baud rate (16 times faster or more is common). At 2400 baud each bit is present for:

$$\frac{1}{2400} = 0.417 \text{ ms.}$$

and so the receiver should be capable of reading the input at least every 26 μs (0.417ms/16). Initially the data line is at logical 1 and the receiver responds as follows:

Step 1   The receiver keeps reading the input port until it detects that the signal line has gone low.

Step 2   It then waits for half the bit time (0.209 ms) and checks that the line is still low. If it is, it regards this as a start bit. If the line is high, it is assumed that the logical 0 was just noise on the line and the receiver must return to step (1).

Step 3   The delay at Step 2 effectively gives a reference point which is in the middle of the time when a bit is being transmitted. The receiver now delays for 0.417 ms (1 bit time) before sampling the data line again. This delay places it in the middle of the time when the first data bit is present. The data on the data line is then read and stored for serial to parallel conversion.

Step 4   Step 3 is repeated for the remaining data bits. When all 8 bits have been read the serial to parallel conversion is completed and the data byte stored for future use.

Step 6   The receiver has two more delays of 0.417 ms and each time it checks that the data line is high (the two stop bits). If it is not, an error is reported.

Step 7   If more data is to be received then, after the last stop bit is received, the receiver once again starts sampling the data line at high speed looking for the next start bit.

At first sight Step 7 may seem a little unusual given that the baud rate and data format is known. However, manufacturing tolerances mean that it is highly probable that the receiver and transmitter are not working at exactly the same speed. Looking for the next start bit gives the receiver a chance to correct any cumulative error and ensure that it is once again sampling the data line near the center of each data bit.

   Two way communication may be achieved with the arrangement above by transmitting a predetermined data byte to indicate when the direction of data flow is to be reversed. Using a standard I/O port and software for serial communication in the manner described above is sometimes referred to as a **software UART**. In small dedicated systems, or when adapting existing systems for serial communication, the software UART may be the most economical method. However, the software approach uses a lot of processor time and, if speed is a problem, then a hardware solution may be more appropriate.

Now complete Exercise 11.1 at the end of this chapter.

11.2 HANDSHAKING

This technique is often used where more than one byte of data needs to be transferred and there is a disparity between the speed of operation of the transmitter and receiver (e.g. parallel data being sent to a printer or serial data being sent to a modem for transmission over a telephone line). The basic process consists of:

    (1)   Transmitter places a byte of data in the output register of the appropriate I/O device and then sends a signal to the receiver to indicate that data is available.

    (2)   When it is free to do so, the receiver reads the data and signals the transmitter that it may now send the next byte of data.

11.2.1 Parallel Communication with Handshaking

The basic configuration for parallel communication with handshaking is shown in Figure 11.3.

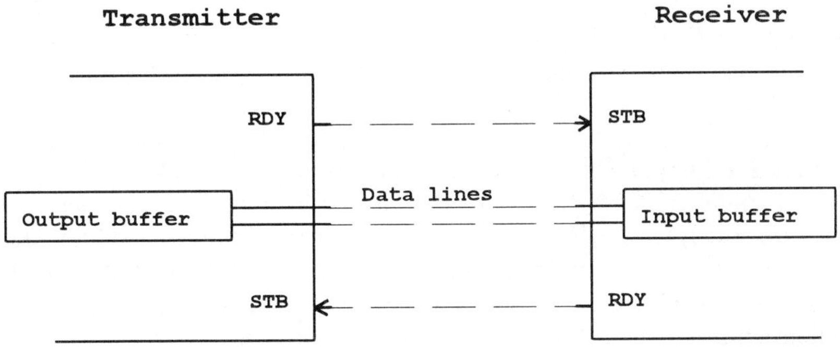

```
        Transmitter                              Receiver

              ┌─────────────┐           ┌─────────────┐
        RDY   ├─ ─ ─ ─ ─ ─ ─ ─ ─ ─ ─ ─►│ STB
              │             │           │
              │             │  Data lines │
  ┌────────────┐            │  ─ ─ ─ ─ ─  │   ┌────────────┐
  │Output buffer│───────────│─ ─ ─ ─ ─ ─ ─│───│Input buffer │
  └────────────┘            │  ─ ─ ─ ─ ─  │   └────────────┘
              │             │           │
        STB   │◄─ ─ ─ ─ ─ ─ ─ ─ ─ ─ ─ ─┤ RDY
              └─────────────┘           └─────────────┘
```

Figure 11.3

The sequence becomes:

Transmitter	**Receiver**
(1) Check STB to make sure that receiver is ready for data.	
(2) Put new data in output buffer and activate RDY.	
(3) Either loop or proceed to another activity.	
	(4) On receipt of STB, deactivate RDY.
	(5) Complete any essential current task.
	(6) Read data and put RDY active.
(7) On receipt of STB, deactivate RDY	
(8) If more data is to be sent, complete any essential current task and return to (2)	

The names of the signals involved will depend upon the I.C's being used e.g.

Transmitter Signals	Figure 11.1	Z80-PIO	8255 PPI*
Output data ready	RDY	A RDY	OBF_A
		B RDY	OBF_B
Receiver ready	STB	A STB	ACK_A
for next byte		B STB	ACK_B

Receiver Signals	Figure 11.1	Z80-PIO	8255 PPI*
Data waiting to	STB	A STB	STB_A
be read		B STB	STB_B
Ready for next byte	RDY	A RDY	IBF_A
		B RDY	IBF_B

* When used in a handshaking mode the 8255 uses Port C for these control lines and so you will not find them on a pinout diagram but you will find them elsewhere on an 8255 data sheet.

11.2.2 Serial Communication with Handshaking

There are is number of different approaches in common use for serial data transfer - even the very widely used RS232 standard has a number of different configurations. To describe them all would be outside the scope of this book. Instead we shall refer to one variant of the RS232 configurations which is often used when a microprocessor is controlling hardware which is situated some distance from the processor. The basic configuration is shown in Figure 11.4

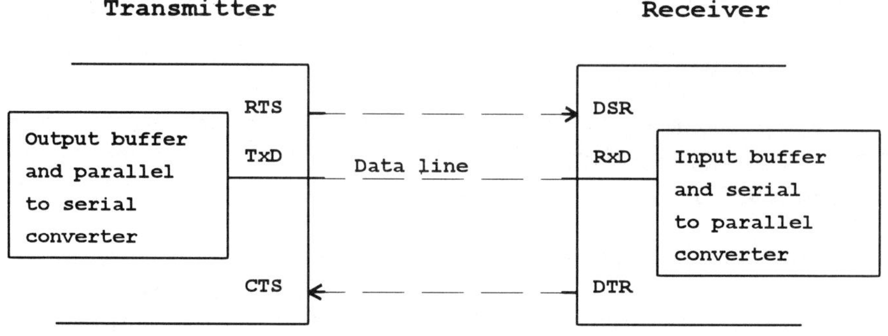

Figure 11.4

The operation of this is very similar to parallel communication:

    (1)   Transmitter checks CTS (**Clear To Send**) to ensure that the receiver is ready for data.

    (2)   Transmitter sends RTS (**Request To Send**) to DSR (**Data Set Ready**) and sends data.

(3)   Receiver deactivates DTR (**Data   Terminal   R**eady) indicating that it
          is not yet ready to receive more data.

(4)   Receiver reads data and reactivates DTR.

11.2.3 UARTs

UARTs (**U**niversal   **A**synchronous   **R**eceiver/**T**ransmitter) have been developed for
serial communication in a variety of different digital systems.   Some were developed
specifically for microprocessor applications and provide additional features which make
them particularly suitable for this environment. The Motorola MC6850 ACIA
(**A**synchronous   **C**ommunications   **I**nterface   **A**dapter), is an example of such a device.
It may be used for both one way and two way communication with or without hand
shaking.

The MC6850 may be regarded as having *four* registers (control register, status
register, transmit data register and receive data register) at *two* addresses which are
usually consecutive.   The control register and transmit data register are write only
registers whereas the status register and receive data register are read only registers.
Internal logic decodes the signal arriving at the register select pin (usually driven from
address line A0) and the signal on the Read/Write pin to determine which of the four
registers is to be accessed.

When transmitting data the first step is to send a control word to the control register
indicating the format of the data (number of data bits, number of stop bits etc.). Once
this has been done a byte of data may be written to the transmit data register. The
ACIA then transmits this data independently of the processor. Whilst this is taking
place the processor may either continue with other operations until an interrupt is
received or it may sit in a loop reading the status register waiting until the contents of
the status register indicate that the data has been read by the receiver. When the
receiving device has read the data it will indicate this to the ACIA, the status register
will be adjusted and an interrupt may be generated (see section 11.4.2).

Reception of data follows a similar pattern. A control word is written to the control
register indicating the form of the data. The processor then reads the status register to
see if there is valid data in the receive data register. If data has been received it may be
read. The processor may again either continue with other operations until an interrupt
is generated or it may sit in a loop reading the status register.

Now complete Exercise 11.2 at the end of this chapter.

11.3 INTERFACING WITH ANALOGUE DEVICES

In all our work so far we have been assuming that external devices are digital in nature
and we may communicate with them using 0's and 1's. Many of the pieces of
equipment to which a microprocessor might be connected are however analogue in
nature. A temperature sensor for example usually indicates temperature by a voltage
which may take any value between the minimum and maximum of its range, speed
control of small dc motors is often achieved by altering the voltage at its terminals etc..

11.3.1 Digital Representation of Analogue Signals

The temperature of the room in which you are sitting is probably somewhere between 20°C and 24°C but could take any value between. To represent this with a digital signal presents some difficulty and we must resort to coding. If we used a single line (one bit) we might decide to represent any temperature below 22°C by logical 0 and any temperature of 22°C and above by logical 1. This has some parallels with a room thermostat. Using a single bit is rather inaccurate since the best you can do is to say that the room is either above or below a given temperature. If we use two bits there are four combinations to chose from. These could be used to code the temperature as follows:

> 00   < 21
> 01   between 21 and 22
> 10   between 22 and 23
> 11   > 23

In mid range we can establish the temperature to within one degree. If this was not sufficiently accurate we could try three bits. In mid range this would enable us to get to within half of one degree. The more bits we use, the greater the accuracy, but we can never guarantee to be able represent an analogue signal with 100% accuracy. Figure 11.5 for example shows a voltage signal which rises from 0V to 10V at a uniform rate and along side it is a diagram showing the best approximation that is possible with a 3 bit binary signal.

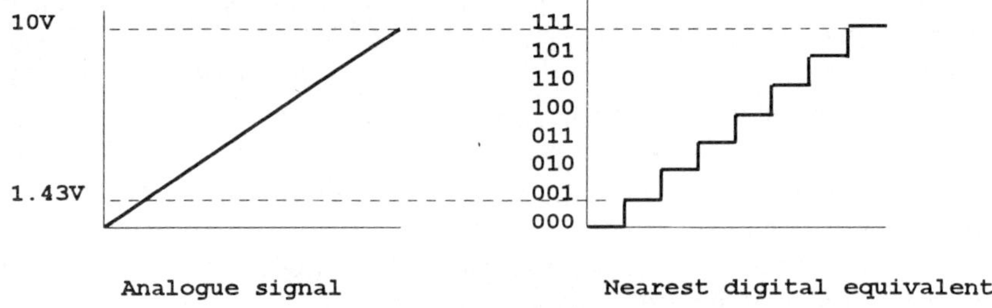

Figure 11.5

Notice that, with eight different values for the digital signal, the original signal is split into seven steps. Each step in the digital signal represents a change of 10V/7 ≏ 1.43V in the analogue signal. In other words, the best accuracy we can achieve is 14.3%. More correctly we should say that we can achieve a **resolution** of 14.3% since accuracy does not depend solely on resolution. In general, resolution is given by:

$$\text{resolution} = \frac{\text{analogue range}}{2^n - 1} \qquad \text{where } n = \text{number of bits}$$

With more bits, greater resolution is possible. I.Cs which use either 8 or 12 bits are readily available for converting between digital and analogue signals. With these devices the resolutions possible, expressed as a percentage, are:

8 bit $\qquad\qquad \frac{1}{2^8 - 1} \times 100 \triangleq 0.4\%$

12 bit $\qquad\qquad \frac{1}{2^{12} - 1} \times 100 \triangleq 0.02\%$

11.3.2 Digital to Analogue Conversion

When a microprocessor based system needs to send an analogue signal, a **DAC** (Digital to Analogue Converter) I.C may be used. These devices are usually connected to an output port as shown in Figure 11.6. Therefore, from a programmers point of view, all that is necessary to send an analogue signal is to send the equivalent digital signal to the address of the output port to which the DAC is connected.

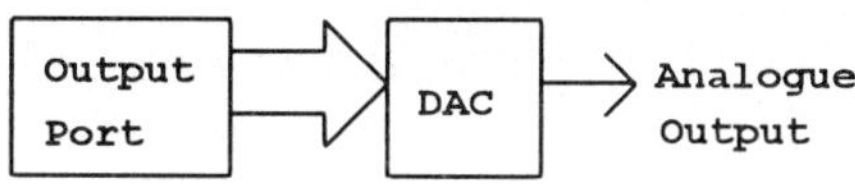

Figure 11.6

DACs can convert signal quite rapidly (typically 1 µs or faster). This means that, for many applications, the conversion may be regarded as instantaneous. There are however a number of other problems that need to be addressed e.g.

(a)   The amount of current that may be taken from a typical DAC is relatively small.

(b)   DACs need an analogue reference voltage on which to base the conversion.

Full analysis of the requirements for adding a DAC to a microprocessor based system are outside the scope of this book. However you may like to experiment with the circuit shown in Figure 11.7 which uses a DAC0800.

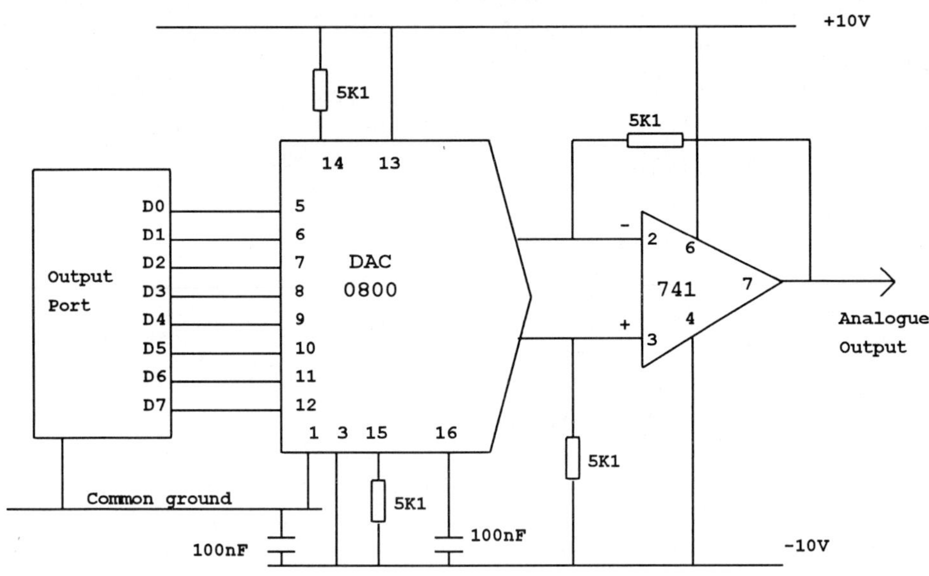

Figure 11.7

11.3.2 Analogue to Digital Conversion

There are many types of **ADCs** (Analogue to Digital Converters) each with its own merits and deficiencies. At one extreme we have the **flash converters**. These can convert at extremely high speeds (tens of megahertz) and are frequently used in video applications. They tend to be rather expensive and so, unless their high speed is essential, other types of converter may be preferable. At the other extreme we have the **ramp converters** and **successive approximation converters**.

Both the ramp and the successive approximation ADCs start with a digital signal and use a DAC to convert the digital signal to an analogue signal. This signal and the original analogue input signal are then passed to a comparitor which indicates if the output of the DAC is greater than or less than the original analogue signal. On the next clock pulse the input to the DAC is adjusted and a new comparison made. This clearly takes a significant length of time and these types of ADCs have a minimum of three inputs in addition to the analogue signal: **SOC** (Start **Of** Conversion), **EOC** (End **Of** Conversion) and **Ck** (Clock). The basic configuration for using an ADC with a microprocessor based system is shown in Figure 11.8.

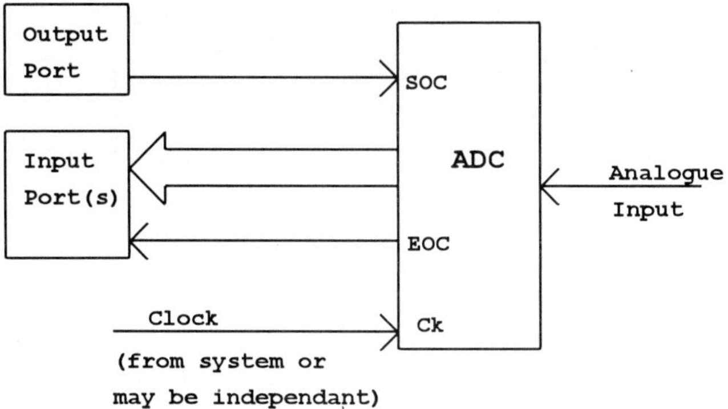

Figure 11.8

The software required to use this sort of ADC is a little more involved than the DAC. the main steps are shown in Figure 11.9.

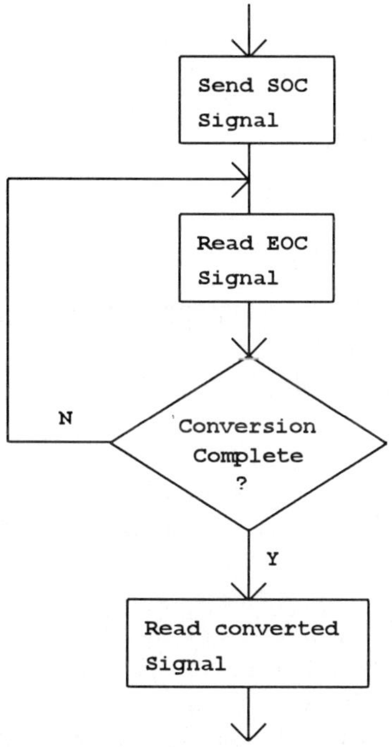

Figure 11.10

Now complete Exercises 11.3 and 11.4 at the end of this chapter.

11.4 MULTIPLE DEVICE INTERFACING

Although relatively slow in terms of electronic equipment, the microprocessor operates at speeds which are far faster than can be achieved by any mechanical equipment or by humans. This means that a single microprocessor can be used to control a number of external devices and/or respond to many actions initiated by people. When controlling multiple devices one of two techniques is usually used - **polling** and **external interrupts**.

11.4.1 Polling

Polling is a communication which is initiated by the processor. The technique involves the processor going to each device in turn and checking to see if any action is required. If it is, the appropriate action is taken and the processor moves on to the next device. Consider, for example, an arrangement in which a microprocessor based system is to send serial data to one device (A), receive serial data from another device (B) and continuously logg the value being presented via a successive approximation ADC from an analogue device (C). The format of the program might be:

    (1)   send SOC to device C
    (2)   read CTS from device A
            if clear to send
                    load next byte of data and activate RTS
    (3)   read DSR from device B
            if data set ready
                    read next byte from device B and activate DTR
    (4)   read EOC from device C
            if conversion complete
                    read converted data
                    log result in table or on disk
                    send SOC
    (5)   go back to (2) and repeat

At step (4) notice how, if a device (the ADC in this case) has completed its current task, it is given its next task before the processor moves on to the next device. If a device has not completed its current task the processor simply moves on the next device.

11.4.2 External Interrupts

An interrupt is a communication initiated by an external device. In many respects the processor's response to an interrupt is rather like a call to a subroutine. The principle difference is the way in which the program arrives at the subroutine. With ordinary subroutines, the processor arrives at the start of the subroutine in response to a software instruction such as CALL or JSR. With an interrupt, the processor arrives at the start of the 'subroutine' (now referred to as an **interrupt service routine** or **ISR**) in response to an external signal arriving at one of its pins. The provision of such pins means that external devices can interrupt the current activity of the processor and demand attention.

The number and organization of interrupt pins varies from one processor to another but most will have at least two interrupt pins. On many processors (including Z80 and 8086) one of these is a **Non-Maskable Interrupt** pin which is usually designated **NMI**. Non-maskable means that the processor cannot ignore signals on this pin and it must jump to the interrupt service routine as soon as it completes the current instruction. NMI is used for high priority events.

The second interrupt pin is an **INT**errupt **R**equest pin (designated **INT** on the Z80 and **INTR** on the 8086). Interrupt request pins are sometimes referred to as maskable interrupts which means that the processor may be programmed to ignore them. This may be done under program control by executing a 'disable interrupt' instruction. However, when an NMI occurs, interrupt requests are automatically disabled. This ensures that NMI will not be interrupted by an interrupt request.

The discussion so far has left one important question unanswered - what stops an interrupt from interrupting itself? In the case of NMI the pin is edge triggered and usually negative edge triggered. This means that an interrupt will only occur when the pin is taken from logical 1 to logical 0. To generate a second interrupt NMI must first be taken back to logical 1. The hardware built around devices which have been given access to NMI must ensure that this will not occur until the cause of the interrupt has been removed. In the case of interrupt request, interrupts are automatically disabled when the processor start to respond. With most processors it is the programmer's responsibility to ensure that they are re-enabled at the end of the ISR.

The 68000 has a different approach. Three pins designated IPL2, IPL1 and IPL0 are used to both create and indicate the priority level of an interrupt. 000 on these pins indicates no interrupt. 001 is an interrupt with the lowest level of priority and 111 is an interrupt with the highest level of priority. If there are three or less devices which are to be allowed to generate interrupts then they may each be allocated a pin according to their priority. The pins IPL2, IPL1 and IPL0 may however be driven from an 8 to 3 line encoder. Given that 000 means no interrupt, this will enable up to seven devices to generate interrupts.

As with subroutines the processor will save the return address on the stack before jumping to an ISR. In addition, it may also save other items on the stack and make changes to internal registers which are beyond the direct control of the programmer. To make an orderly return to the point in the program where the interrupt occurred it is important to use the correct return instruction. These are:

Z80   RETI Return from interrupt request
           RETN   Return from non-maskable interrupt

8086   RET Return from interrupt (maskable or non-maskable). IRET is
           also used for software interrupts but these are beyond the scope of
           this book.

68000   RTE   Return from exception.   The 68000 has a number of operations which resemble interrupts most of which are beyond the scope of this book. They are collectively referred to as exceptions and an interrupt is just one form of exception

11.4.3 Mixed Polling and External Interrupts

It is quite common to have a microprocessor based system in which some of its external devices are polled and some generate interrupts when necessary. It might be, for example, that the analogue sensor mentioned in section 11.4.1 was monitoring room temperature and this only needed to be recorded every five minutes. If this were the case we could use a timer I.C. to generate interrupts at five minute intervals and the format of the program would become:

Main Program
(1)   enable interrupts
(2)   read CTS from device A
         if clear to send
                 load next byte of data and activate RTS
(3)   read DSR from device B
         if data set ready
                 read next byte from device B and activate DTR
(4)   go back to (2) and repeat

Interrupt Service Routine
(1)   send SOC
(2)   do
         {
                 read EOC
         } while conversion not complete
(3)   read converted signal and log result
(4)   return from interrupt

11.5 SETTING UP FOR INTERRUPTS

When an interrupt occurs the processor must know where to go to start executing the interrupt service routine. The technique for establishing this address will depend upon the processor and the way it was initialized. We must therefore look at each processor in turn.

Z80   **NMI** When an NMI occurs the Z80 will jump to address 0066 hex and start executing code from there. It is not essential that the main body of the ISR starts at this address. Indeed, in many systems, an unconditional jump instruction is placed at this address.

INT   There are three ways in which a Z80 might be set up to respond to INT. These are known as mode 0, mode 1 and mode 2 and are selected by issuing one of the instructions IM0 or IM1 or IM2.

Mode 1 is the simplest to use and is similar to NMI. When set for mode 1 interrupts, an interrupt request will cause the processor to jump to address 0038 hex and start executing code.

In mode 0 the processor looks to the device which has caused the interrupt to provide a single byte instruction - this is usually one of the restart (RST) instructions.

Mode 2 is the most difficult to use. After receiving a mode 2 interrupt the processor looks to the interrupting device to provide the low byte of a 16 bit address. Earlier programming, such as the cold start procedures, must have placed the high byte of this address in the **I** register (Interrupt vector register). The address formed by these two bytes is not the start address of the ISR but indicates where the start address is to be found. This rather complex arrangement gives the Z80 the capability of giving an individual response to up to 128 different interrupting devices.

8086   Both NMI and INTR get the address of the ISR from a table of addresses in memory known as the **Interrupt Vector Table**. This resides in the first 1K of memory. Each vector table entry is four bytes long and contains the offset segment address for the ISR. (You should be aware that many texts on the 80x86 family refer to an ISR as an **interrupt handler**.)

When an interrupt occurs the processor looks to the interrupting device to provide an 8-bit interrupt number. The processor multiplies this number by four to get the address in the interrupt vector table where the address of the ISR is stored. Interrupt numbers 0 to 4 have specific uses:

No. 0   Divide by zero error.

No. 1   Single step.

No. 2   Entry point for NMI (i.e. interrupt number 2 has the same effect as NMI).

No. 3   Break point.

No. 4   Interrupt on overflow.

Interrupt numbers 5 to 31 are reserved for use by Intel but numbers 32 to 255 are free. Note however that manufacturers of operating systems may have used some of these interrupt numbers and it is important to check the manual before using any interrupt.

The reader should not confuse an interrupt request using INTR as described above with IRQs which are often referred to in documentation relating to the desktop PC. IRQs are associated with an interrupt controller chip such as the 8259. In its simplest configuration the 8259 may receive interrupt requests from up to eight different devices. It will prioritize these requests, send a signal to INTR on the 8086 and, when asked to do so, will send the appropriate interrupt number.   From this is should be apparent that part of the system initialization processes must be to program the 8259 with the appropriate interrupt numbers.

68000   The 68000 has a number of modes of operation, referred to as **exceptions**, which result in the normal execution of a program being interrupted.   All of them select the address of the ISR from a vector table referred to in the data sheet as an **exception vector table**. This table occupies a 1K block of memory and each vector address is 4 bytes long. For external interrupts of the type we have been discussing, the signals on pins IPL2, IPL1 and IPL0 direct the processor to get the address of the ISR from an area of this table allocated to **interrupt autovectors**. The table is located in an area of memory which is not normally accessible to the user. However, if you are using a 68000 training system, it may be that the manufacturer of the system has placed a vector in this table which directs the processor to an area of user RAM.

11.6 Timers

The passing of time is analogue in nature and thus presents problems for time dependent events in microprocessor based systems. Software solutions such as those given in section 10.2 are frequently used but these hog processor time and solutions involving the use of some sort of timer circuit may be more appropriate. In choosing a timer circuit there are two extremes to consider:

      (a)   do we want the timer to run continuously so that we may interrogate it occasionally to establish how much time has passed since it was started?

      (b)   do we want the timer to run only when instructed to do so and give a *time out* signal when a certain period of time has elapsed?

If we want the timer to run continuously a simple solution might be to use a basic counter IC which is connected to the data bus via tristate buffers as shown in Figure 11.10

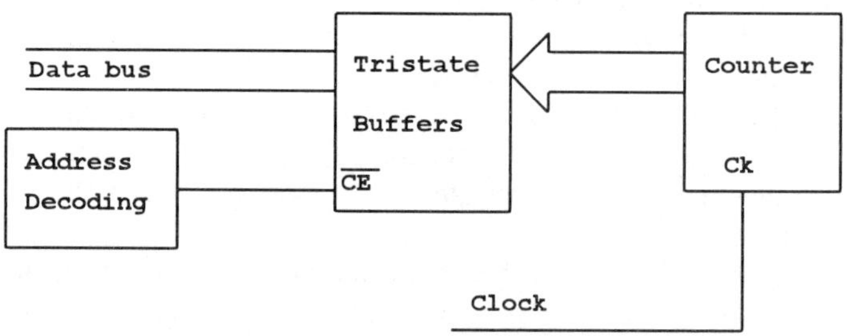

Figure 11.10

If the circuit providing the clock pulse is designed to have a convenient frequency such as 1 kHz then the counter will indicate the passing of time in milliseconds.

When a *time out* signal is required in a microprocessor based system the most usual approach is to use a timer IC.   In spite of their name, most timer ICs are, in fact, counters.   A register is loaded with the required value and then decremented at regular intervals.   When the counter reaches zero a time out signal is generated.   The time out signal is usually an interrupt but the timer might also be indicated by other pins on the IC.   The Z80-CTC (Z80 **C**ounter / **T**imer **C**ircuit), for example, has four channels that may be used for timing.   Three of these have ZC/TO pins which may be used to send signals to other devices.   All four channels can generate an interrupt by putting pin 13 (designated INT) low.

Now complete Exercise 11.5 at the end of this chapter.

Exercise 11.1   *** *suggested for portfolio* ***
Refer to   Case   Study 1.   Write a program to implement routine (d).   The program should be written as a stand alone item and you should ignore the initial need for the switches to be at 11.

Exercise 11.2   *** *suggested for portfolio* ***
Write a report which examines the communication methods which might be used between a computer and a robot for the situation described below.   The report is to be in the form of a discussion document which indicates the pros and cons of each method and then places in order of preference.

> It is proposed to use messenger robots to transport components and equipment around a large industrial site.   The robots resemble motorized super market trolleys and each has a built-in microprocessor based control system.   When a robot arrives at a docking point it automatically engages with a connector. This connector provides a link to a master computer which passes information to the robot's microprocessor based system indicating its next destination and the route it should follow.   When the robot sets off it will be following tracks marked out on the floor.   The information it received from the master computer indicates at which junctions of these tracks it should turn.

Exercise 11.3
Figure 11.11 shows a circuit which may be attached to the output port of a microprocessor based system and, together with suitable software, may be used for waveform synthesis.   Write a program which will use this circuit to generate a triangular wave which rises to a maximum in 1 second and falls back in 0.1 seconds.

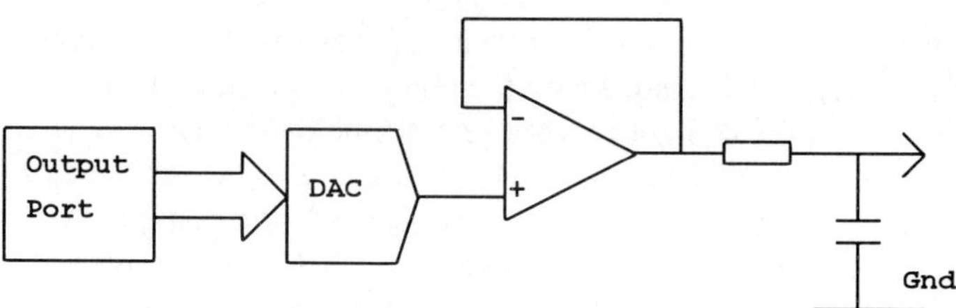

Figure 11.11

Exercise 11.4 *** suggested for portfolio ***

The pressure inside a pressure vessel is to be monitored by a microprocessor based system and the relevant parts of the specification are:

(a)   range of outputs from the pressure sensor 0 -10V linear.

(b)   there is a need to record the pressure at 1 second intervals.

(c)   maximum rate of change of pressure is 10% per second

(d)   an over pressure alarm signal is required if the output from the pressure sensor is 8V +/- 0.1Vor greater.

(e)   an under pressure alarm signal is required if the output from the pressure sensor is 2.5V +/- 0.1Vor less.

Write a short report indicating the characteristics you would need from an ADC to meet these requirements. In your report include reference to:

(a)   the type of ADC to be used (ramp, flash etc.).

(b)   the number of bits the ADC will have and how many bits will be used.

(c)   how the SOC and EOC signals are to be handled.

Exercise 11.5 *** suggested for portfolio ***

An environment conditioning system for an office block is to have a microprocessor based controller and provide the following functions:

(a)   monitor both internal and external temperature and humidity.

(b)   respond to changes in the external environment and make preemptive adjustments to the internal conditions so that changes in temperature and humidity are not noticed by personnel.

(c)   respond to fire sensors by sounding an alarm, operating a sprinkler system in the appropriate area, flashing exit route signs, telephone the emergency services and play a prerecorded message.

Write a short report indicating how this might be implemented. In your report include reference to:

(a)   the parts of the system where polling is most appropriate and where external interrupts would be used.

(b)   the parts of the system where serial data transfer is most appropriate and where parallel data transfer is most appropriate.

(c)   the parts of the system which would require either ADC or DAC.

(d)   the parts of the system where the use of a timer would be appropriate.

Appendix A - Case Study Scenarios

CASE STUDY 1

A training establishment believes there is a gap in the market for a course giving a good all round introduction to microprocessors. Having decided that a separate microprocessor board is necessary it is not sure which processor to use and you have been asked to make a recommendation. In addition to providing its own training facilities it also wishes to franchise the course to schools and colleges.

Your recommendations for a processor have been accepted and it has been decided that an I/O board is required. This is to consist of 8 DIL switches for input and 8 LEDs for output. You have been asked to write a sales demonstration program which will use the initial settings of the switches at bits 7 and 6 to select any one of the four routines (a - d) given below. After reading the switches the program shall wait for 2 seconds before commencing the selected routine to allow changes to be made to the switches if necessary.

a) Switches at 00. The program should read an 8 bit NBCD number from the switches, add 55 decimal to the value and display the result on the 8 LEDs in natural binary coded decimal. If the result is greater than 99 decimal all LEDs are to be illuminated.

b) Switches at 01. LED 0 should flash at a rate determined by bits 7 and 6 of the switches as shown below. After each flash the state of the switches should be checked to see if the flash rate has changed.

00	4	Hz
01	2	Hz
10	1	Hz
11	0.5	Hz

c) Switches at 10. Create an effect whereby a single illuminated LED appears to run along the LEDs from left to right (known as a running light display). When the illuminated LED reaches the right hand end it is to extinguish and the process repeated.

d) Switches at 11. Asynchronous reception of serial data is to be simulated. Serial data is to be entered manually using switch 7. On receipt of a start bit all LEDs are to be extinguished. Then, as the serial data is received, it is to be progressively displayed on the LEDs. When a stop bit is received the program shall wait for the next start bit before repeating the process. The transmission format is to be:

Baud rate	0.5	(far slower than any standard baud rate but this is necessary for manual entry of data)
Date bits	8	
Start bits	1	
Stop bits	1	

CASE STUDY 2

The main parts of a microprocessor based intruder alarm for domestic purposes are shown in Figure A.1. The requirements of this device are:

1. Arming and disarming of the system is to be by means of a key operated master switch.

2. There are to be three circuits of sensors. Circuit 1 is to be the entry and exit point and shall consist of one sensor only. Circuits 2 and 3 may consist of any number of sensors.

3. Whatever type of sensors are used they will respond as though they are a normally closed switch until activated by an intruder.

4. Circuit 2 and/or circuit 3 may be disabled by turning the master switch to the disable position and holding it there whilst the appropriate disable switch is pressed. The master switch is then returned to the 'On' position. Once the master switch is in the 'On' position changes to the disable switches have no effect on the settings.

5. The siren has its own power supply but uses a signal from the control box to prevent it sounding. If this signal is removed either by the control box or by someone tampering with the wiring the alarm is to sound.

6. When first enabled, 20 seconds is to be allowed before the system becomes active.

7. If the sensor of circuit 1 is opened the person entering is to have 20 seconds in which to switch off the alarm.

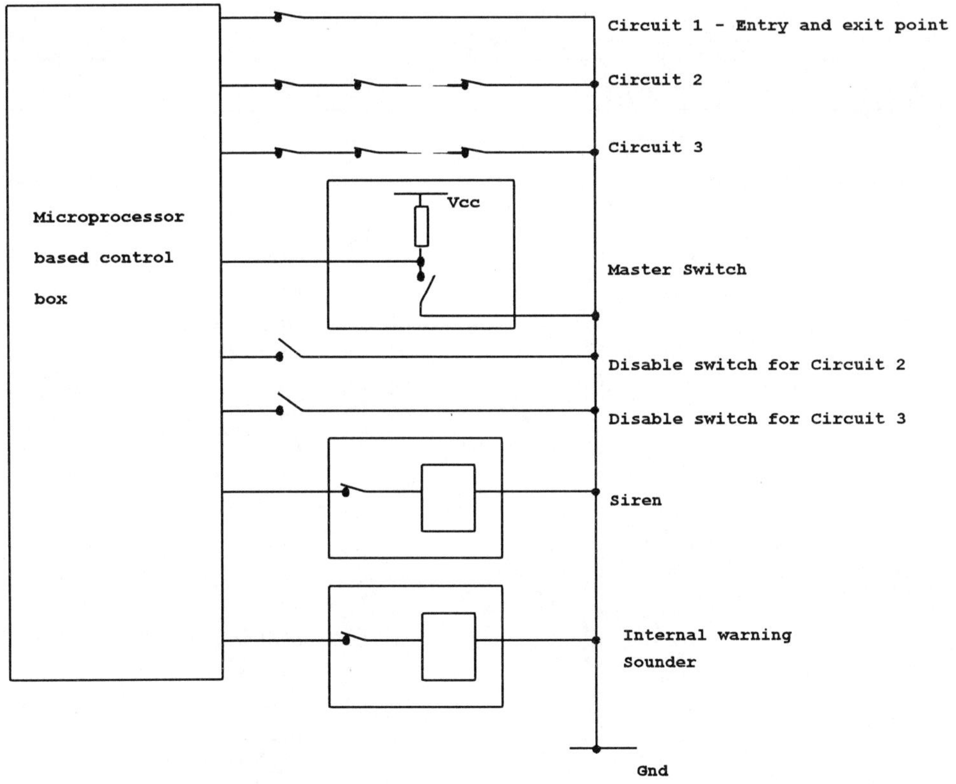

CASE STUDY 3

A security system is required for a large industrial complex. The control unit is to monitor closed circuit television cameras as well as a many other sensors. If an intruder is detected an alarm is to sound in the security bay and it must be possible to display the signal from any of the television cameras on a monitor similar to that on a desktop computer. It must also be possible for the security officer to receive urgent messages as a caption on the screen.

The customer has specified that the system must be easy to use and he anticipates that you will be using an operating system similar to that found on desktop computers.

Appendix B: Z80

B.1 Pinout Diagram

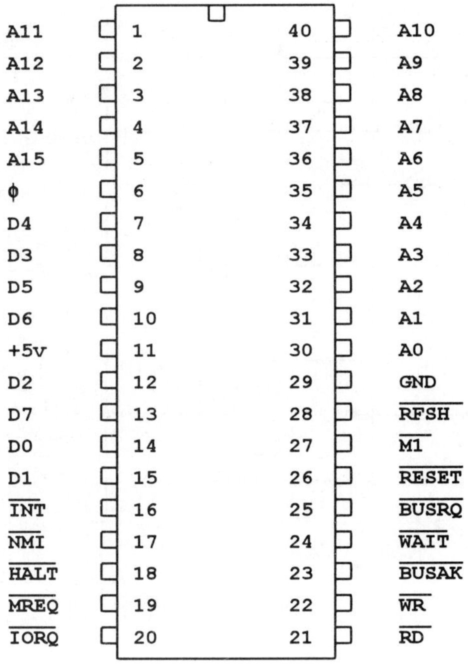

Left	Pin		Pin	Right
A11	1		40	A10
A12	2		39	A9
A13	3		38	A8
A14	4		37	A7
A15	5		36	A6
ϕ	6		35	A5
D4	7		34	A4
D3	8		33	A3
D5	9		32	A2
D6	10		31	A1
+5v	11		30	A0
D2	12		29	GND
D7	13		28	$\overline{\text{RFSH}}$
D0	14		27	$\overline{\text{M1}}$
D1	15		26	$\overline{\text{RESET}}$
$\overline{\text{INT}}$	16		25	$\overline{\text{BUSRQ}}$
$\overline{\text{NMI}}$	17		24	$\overline{\text{WAIT}}$
$\overline{\text{HALT}}$	18		23	$\overline{\text{BUSAK}}$
$\overline{\text{MREQ}}$	19		22	$\overline{\text{WR}}$
$\overline{\text{IORQ}}$	20		21	$\overline{\text{RD}}$

Reproduced by permission. Copyright 1980 Zilog Inc.

B.2   Internal Structure

The diagram below shows a model for the internal structure of a Z80 microprocessor which was developed for teaching purposes by the author. This conceptual model was developed from an original diagram published by Zilog and from works by other authors. No claim for originality is made.

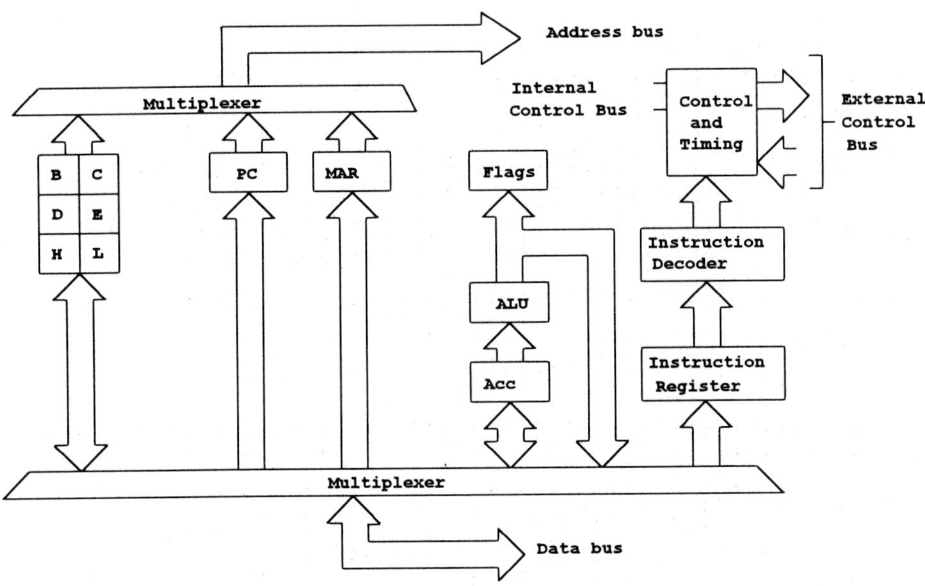

Appendix C: 8086

The diagrams and notes in this appendix are based primarily on the 8086. However, *at the level of this book*, they work well for other members of the 80x86 family. If readers wish to work at much greater depth they should obtain the appropriate data sheet from Intel Corporation

C.1 PINOUT DIAGRAM

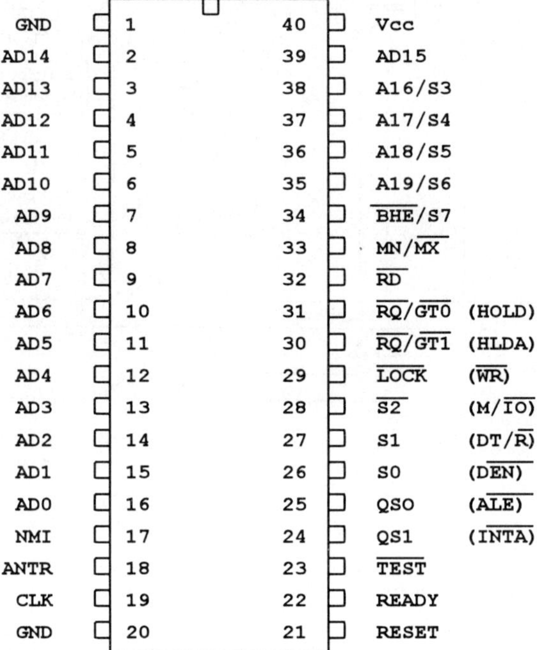

Reprinted by permission of Intel Corporation. Copyright 1969 Intel Corporation

C.2 INTERNAL STRUCTURE

The diagram below shows a model for the internal structure of an 8086 microprocessor which was developed for teaching purposes by the author. This conceptual model was developed from an original diagram published by Intel Corporation and from works by other authors. No claim for originality is made.

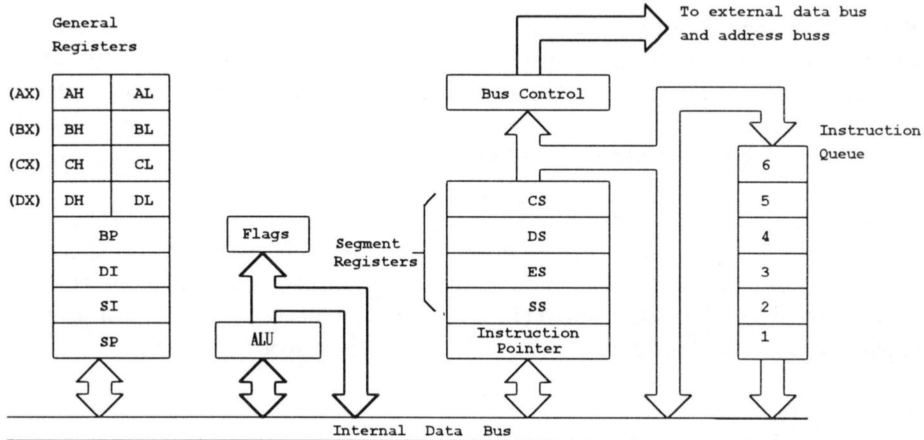

C.3 MEMORY SEGMENTATION

If you are to be using an 80x86 processor for your programming exercises you should read the following notes before starting Chapter 7. Having completed Chapter 7 you should read through these notes again and be sure that you understand how the processor generates a 20-bit address from two 16-bit numbers.

As programming techniques developed, particularly in the 1960s, programmers started to segment their programs and reserved areas of memory for code, areas for data etc.. This was a software approach but, with the advent of the 8088 and 8086 processors, Intel took segmentation a stage further and provided a hardware mechanism for segmenting memory. The registers inside 80x86 family of processors include four 16-bit segment registers designated **CS (Code Segment)**, **DS (Data Segment)**, **SS (Stack Segment)** and **ES (Extra Segment)** specifically for this purpose.

If a programmer wanted to locate his code at addresses starting from say, 10000_{16} and his data at addresses starting from say, 60000_{16} then one of the first tasks of the program would be to load the **most significant 16 bits** of each address into CS and DS respectively. From then onwards memory locations are accessed by using a **16 bit offset or logical address**. Thus if CS = 1000_{16} (note the *four* hexadecimal digits) and there was an instruction to jump to address 1234_{16}

```
e.g.     jmp 1234h
```

Then the destination address for the jump would be:

```
        1000          - contents of CS
   +    1234          - offset
        11234h        - physical address for jump
```

Software tools used for the debugging of 80x86 programs often indicate addresses in terms of a **base address**, which might be one of the segment registers, and an **offset**. The base address and offset are usually separated by a colon to give the general form:

base address:offset

Thus a physical address such as **11234h** might be represented as **1000:1234h**.

By default, the processor will automatically use the CS register to calculate the physical address of any instruction and will use the DS register to calculate the physical address of the data. The reader should be aware that the processor may be instructed to use other registers for base addresses but this is outside the scope of this book.

Example C.1
Calculate the physical addresses represented by 51C8:1A30h and 4453:F180h.

```
    51C8                    4453
    1A30                    F180
    536B0                   536B0
```

Example C.1 above illustrates an important point about segment registers. If we had loaded CS with 4453 and DS with 51C8 then there is a risk that we might be trying to treat the physical address 536B0 as both an item of data and part of a machine code instruction. Fortunately, provided that we observe a few simple rules such as always using labels to refer to addresses and are only writing short programs, we can rely on the assembler and operating system to solve these problems for us.

Appendix D: 68000

D.1 PINOUT DIAGRAM

Left	Pin		Pin	Right
D4	1		64	D5
D3	2		63	D6
D2	3		62	D7
D1	4		61	D8
D0	5		60	D9
$\overline{AS}$	6		59	D10
$\overline{UDS}$	7		58	D11
$\overline{LDS}$	8		57	D12
$R/\overline{W}$	9		56	D13
$\overline{DTACK}$	10		55	D14
$\overline{BG}$	11		54	D15
$\overline{BGACK}$	12		53	GND
$\overline{BR}$	13		52	A23
Vcc	14		51	A22
CLK	15		50	A21
GND	16		49	Vcc
$\overline{HALT}$	17		48	A20
$\overline{RESET}$	18		47	A19
$\overline{VMA}$	19		46	A18
E	20		45	A17
$\overline{VPA}$	21		44	A16
$\overline{BERR}$	22		43	A15
$\overline{IPL2}$	23		42	A14
$\overline{IPL1}$	24		41	A13
$\overline{IPL0}$	25		40	A12
FC2	26		39	A11
FC1	27		38	A10
FC0	28		37	A9
A1	29		36	A8
A2	30		35	A7
A3	31		34	A6
A4	32		33	A5

Reproduced by permission of Motorola Inc. Copyright 1985 Motorola Inc.

D.2 INTERNAL STRUCTURE

The diagram below shows a model for the internal structure of a 68000 microprocessor which was developed for teaching purposes by the author. This conceptual model was developed from an original diagram published by Motorola and from works by other authors. No claim for originality is made.

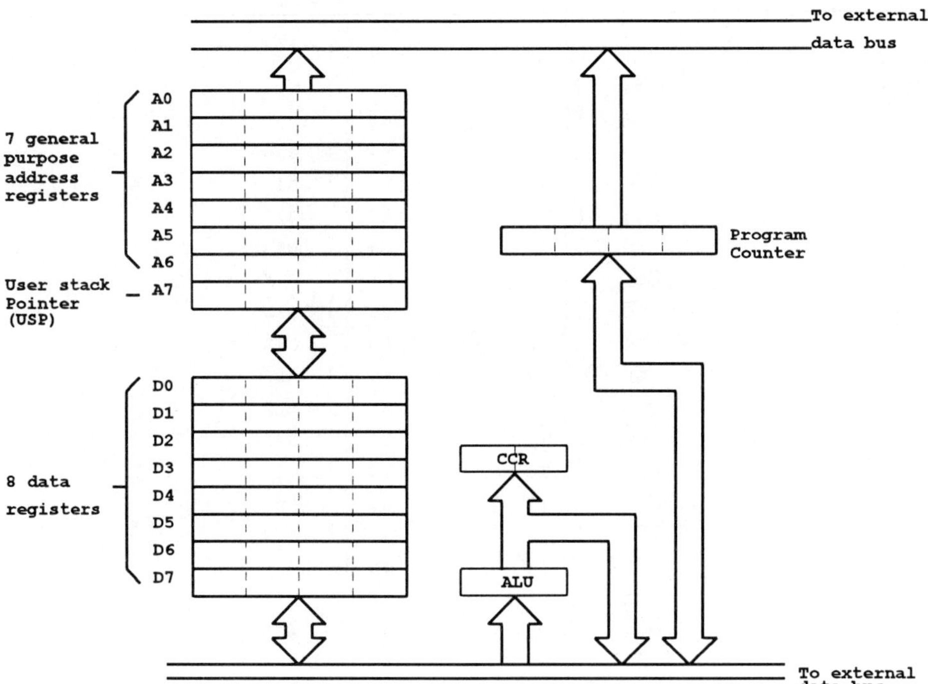

Appendix E - Summary of Selected Processor Characteristics

N.B. The summary of characteristics given below identifies only those characteristics of the device which are relevant to this course. Other features may be available and if more detail is needed the reader should contact the manufacturer.

COP410L Microcontroller - National Semiconductors

Size	4 bit			
Memory	On chip: 512 bytes on chip ROM, 32 bytes on chip RAM			
I/O	On chip: one 8 bit bi-directional port, one4 bit bi-directional port, one 4-bit ouput only port, one serial input port and one serial output port.			
Instruction set	Arithmetic	Add, increment and decrement. Subtraction done by complementing and adding.		
	Logical	OR and XOR.		
	Data transfer	Simple data moves and exchange contents only.		
	Test and branch	Skip instructions if certain bits are zero or if a carry has occurred.		
	Bit manipulation	Individual bits may be set or cleared.		
Speed	Instruction execution time typically 16 µs			
Typical use	Large volume products such as toys and controllers for domestic appliances.			
Cost	Very low if ordered in large quantities.			

PIC16C84 Microcontroller - Microchip Technology

Size	8 bit	
Architecture	Harvard	
Memory	On chip: 1024 x 14bit EEPROM for code memory, 64 bytes EEPROM for data memory and 36 bytes of internal registers for other RAM activities.	
	External: none	
I/O	On chip: 13 I/O pins with individual direction control, timer.	
	External: not relevant	
Instruction set	Arithmetic	Add, subtract, increment and decrement
	Logical	AND, OR and XOR
	Data transfer	Simple data moves only
	Test and branch	Skip next instruction on result of test, unconditional jumps, unconditional subroutine calls
	Bit manipulation	Individual bits may be set, cleared or tested
Speed	Instruction execution time typically 400 ns	
Typical use	Small volume control systems.	
Cost	Low/moderate	

8031 Microcontroller - Intel

Size	8 bit
Architecture	Harvard
Memory	On chip: 128 bytes of RAM (Other processors in this family have up to 4Kbytes of ROM or EPROM and 256 bytes of RAM)
	External: I/O ports may be used to provide address and data bus for up to 64Kbytes of ROM and/or RAM.
I/O	Four 8 bit bi-directional parallel ports, two timers, one serial I/O port

Instruction set

Arithmetic	Add, subtract, multiply, divide, increment and decrement
Logical	AND, OR, XOR (also see bit instructions)
Data transfer	A number of move and exchange instructions.
Test and branch	Conditional and unconditional jump instructions, unconditional subroutine calls.
Bit manipulation	Individual bits may be set, cleared or tested. also bitwise AND, OR and NOT may be performed.

Speed	Instruction execution time typically 1 µs.
Typical use	Control applications of moderate complexity particularly where some analysis of data is required.
Cost	Low/moderate

Z80 Microprocessor - Zilog

Size	8 bit
Architecture	Von Neumann
Memory	On chip: none
	External: Up to 64 Kbytes of external memory which may be any combination of RAM and ROM
I/O	On chip: none
	External: Up to 256 external I/O devices may be addressed with I/O instructions

Instruction set

Arithmetic	Add, subtract, increment, decrement
Logical	AND, OR, XOR
Data transfer	A number of ways of moving both bytes of data and blocks of data.
Test and branch	Conditional and unconditional jumps, conditional and unconditional subroutine calls.
Bit manipulation	Set, clear and test any bit in memory or internal register.

Speed	Instruction execution time typically 2 µs.

Typical use	Control applications of moderate complexity, dedicated word processors.
Cost	Low

8086 Microprocessor - Intel

Size	16 bit
Architecture	Von Neumann, CISC
Memory	External: Up to 1 Mbytes of external memory which may be any combination of RAM and ROM
I/O	External: Up to 65536 external I/O devices may be addressed with I/O instructions
Instruction set	Arithmetic Many variants of add, subtract, multiply, divide, increment and decrement.
	Logical AND, OR, XOR, NOT
	Data transfer Wide range of byte and block move instructions
	Test and branch Wide range of conditional jump instructions, unconditional jumps and subroutine calls.
	Bit manipulation Only by use of logical instructions
Speed	Depends on version but instruction execution time typically between 0.5 and 3 μs.
Typical use	Desk top computers, used with other processors to provide a user friendly front end for complex control applications.
Cost	Low/moderate

(This family of processors contains members with 32 bit data busses which can perform floating point arithmetic and are much faster but they are expensive.)

68000 Microprocessor - Motorola

Size	16-/32 bit
Architecture	Von Neumann, CISC
Memory	External: Up to 16 Mbytes of external memory which may be any combination of RAM and ROM
I/O	All I/O is memory mapped.
Instruction set	Arithmetic Many variants of add, subtract, multiply, divide, increment and decrement.
	Logical AND, OR, XOR, NOT
	Data transfer Wide range of byte and block move instructions
	Test and branch Wide range of conditional jump instructions, unconditional jumps and subroutine calls.
	Bit manipulation Test, set and clear bits in registers or memeory
Speed	Depends on version but instruction execution time typically between 0.5 and 3 μs.

Typical use	Larger control applications, master controller in multi processor systems, desk top computers
Cost	Low/moderate

(This family of processors contains members with 32 bit data busses, can perform floating point arithmetic and are much faster but they are expensive.)

PowerPC 603   Microprocessor - IBM/Motorola

Size	32 bit
Architecture	RISC
Memory	External: Up to4 Gbytes of external memory which may be any combination of RAM and ROM, 8 Kbytes of internal data cashe, 8 Kbytes of internal instruction cashe.
I/O	External: effectively I/O is memory mapped.

Instruction set	Arithmetic	Add, subtract, multiply, divide, increment decrement, floating point arithmetic
	Logical	AND, OR, NAND, NOR, XOR plus
some		combined operations
	Data transfer	Many instructions for moving bytes and blocks of integer and floating point data.
	Test and branch	Wide range of conditional and unconditional jumps, conditional and unconditional subroutine calls.
	Bit manipulation	By use of logical and shift/rotate operations - otherwise limitted.

Speed	Instruction execution time varies but in the region of a few tens of nanoseconds.
Typical use	High power desk top computers, large control applications requiring a lot of data processing.
Cost	Moderate/high

Appendix F: Tristate Buffers and Latches

F.1 TRISTATE BUFFERS

For much of our work in digital electronics we tend to think of only two logic levels - logical 0 and logical 1. However, there is a third logic level known as the **high impedance state** which is extremely important in microprocessor based systems. Devices which can place their outputs in any of these three states are known as **tristate devices** and they are sometimes referred to has having **tristate outputs**. Before proceeding to the operation of a tristate buffer we must first review certain aspects of the operation of a transistor.

In general the state of any transistor will be at, or somewhere between, two extremes. At one extreme the transistor is said to be **saturated**. When saturated the resistance between the transistor's collector and emitter is very low and for many purposes it may be regarded as a short circuit. At the other extreme the transistor is said to be **cut off**. When cut off the resistance between the transistor's collector and emitter is very high and for many purposes it may be regarded as an open circuit. If the circuitry connected to the base of a transistor ensures that the transistor can only switch between these extremes then the transistor will behave as an electronic switch.

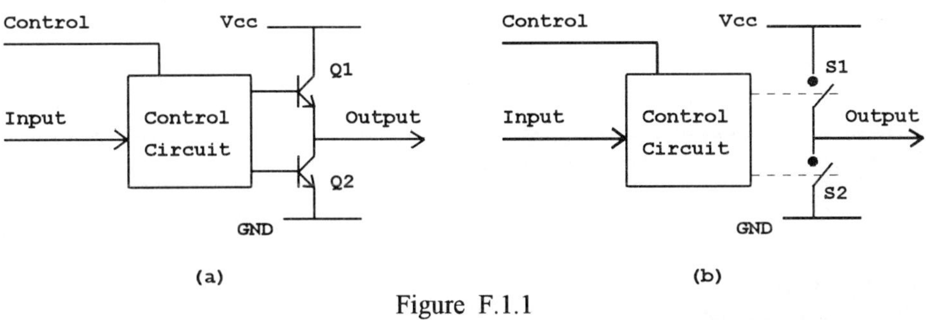

Figure F.1.1

Figure F.1.1(a) shows the main features of a tristate buffer. The control circuit sends signals to the bases of Q1 and Q2 which ensures that these transistors are either saturated or cut off. This means that the transistors will act as switches and a tristate buffer may modeled as shown in Figure F.1.1(b). Thus the operation of a tristate buffer may be described as:

Control	Input	S1	S2	Output
0	0	open	open	high impedance
0	1	open	open	high impedance
1	0	open	closed	0
1	1	closed	open	1

i.e.   if control = 0 then output is high impedance (often abbreviated **hi Z**)

   if control = 1 then output follows input (sometimes referred to as the **active state**)

The tristate buffer provides us with a means of leaving a device physically connected to a circuit but, by sending an appropriate control signal, make the device behave as though it were open circuit.

   Figure F.1.2 shows the circuit symbols for several variants of a tristate buffer together with their truth tables

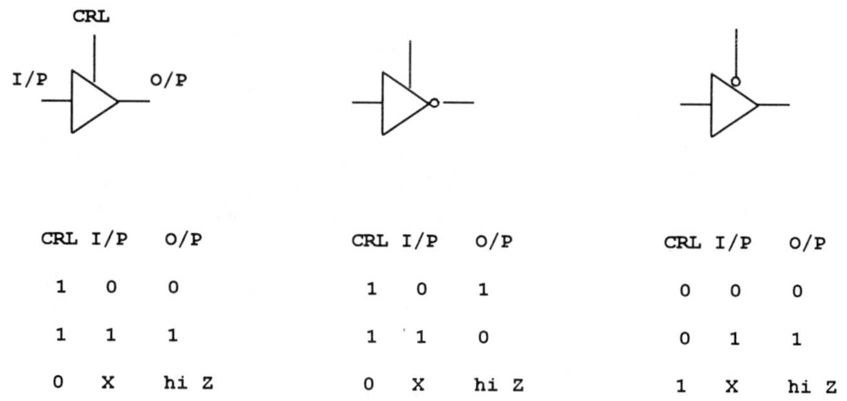

Figure F.1.2

F.2 LATCHES

A latch is a form of memory device. It has a data input line, a data output line and another input which, for historical reasons, is known as a clock input. The simplest form of latch is known as a **transparent D'type latch** and its circuit symbol is shown in Figure F.2.1(a). The operation of this device may be described as:

   when CK = 1   output follows input.

   when CK = 0,   output is held at whatever value it was at just prior to the clock changing to logical 0.

Edge triggered D'type latches are also in common use. Figure F.2.1(b) shows the circuit symbol for a **positive edge triggered D'type latch** and its operation may be described as:

> when CK makes a transition from logical 0 to logical 1 the data at the input is transferred to the output. At all other times the output is held at whatever value it was at when the clock last made an upwards transition.

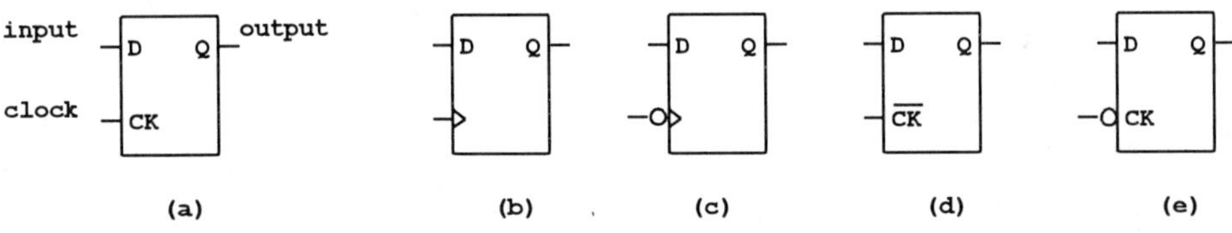

Figure F.2.1

Figure F.2.1(c) shows a negative edge triggered D'type latch. This operates in a similar way to the positive edge triggered device except that latching occurs when the clock input makes a transition from logical 1 to logical 0.

The symbols in Figures F.2.1(d) and F.2.1(e) are both in common use for transparent D'type latches which have a negative clock.

Appendix G: Numbering Systems

Much has been written on the mathematical relationships between different numbering systems. The author has found that, for engineering purposes, a full mathematical treatment causes more problems than it solves. The approach given below has been found satisfactory with students at both this level and higher levels.

G.1 COUNTING IN DIFFERENT BASES

When we talk of a numbering system we are referring to the way in which a set of characters is used to represent a numerical quantity. In normal every day life we use the **denary** numbering system and in computing terms this is often referred to as **decimal** or **base 10**. Because we use base 10 so often, we usually apply the rules for counting with out considering the way in which base ten numbers are constructed. Before moving on to other numbering systems it is worth reviewing the construction of base 10 numbers.

The decimal system (base 10) uses nine characters (1 to 9) and a zero. When we start counting we start in the first column with a number 1 and continue until we get to 9. We have then run out of characters and so, for the next value, we put a 1 in the next column, a 0 in the first column and start again. When we reach 19 we represent the next value by changing the one in the second column to a 2 and changing the 9 in the first column to a zero. The process continues until we reach 99. At this point a one is placed in the third column, the 9s are changed to zeros and so the process continues.

If we wished to count in, say, base 3 the same principles would apply. We would use the characters 0, 1 and 2 and the decimal values 1 to 10 would be represented as 1, 2, 10, 11, 12, 20, 21, 22, 100, 101. When counting in a bases greater than base 10, it is usual to continue after the value 9 with letters. Table G.1 uses upper case letters but lower case are used with equal frequency and it is not uncommon to find text with a mixture of both.

When working at an engineering level with computers and other microprocessor based systems there is a need to be competent with base 10, base 2 (**binary**) and base 16 (**hexadecimal**) numbering systems. If we use the rules above we can draw a table of equivalent numbers:

Table G.1

base 10	base 2	base 16
0	0	0
1	1	1
2	10	2
3	11	3
4	100	4
5	101	5
6	110	6
7	111	7
8	100	8
9	1001	9
10	1010	A
11	1011	B
12	1100	C
13	1101	D
14	1110	E
15	1111	F
16	10000	10
17	10001	11
18	10010	12
19	10011	13
20	10100	14
99	1100011	63
100	1100100	64
255	11111111	FF
256	100000000	100
257	100000001	101

If you examine Table G.1 you will notice that the same combinations of characters can be used to represent different values. The number 11, for example, appears in all three columns but represents different quantities in each - there is clearly a need to distinguish between them. If a number is given with no indication of the base being used then, by default, it is assumed to be base 10. If any other base is used then the base is added in subscript. Thus the quantity we recognize as 19 decimal could be indicated as:

$$19 \qquad 19_{10} \qquad 10011_2 \quad 13_{16}$$

When writing programs for microprocessor based systems subscript is not very convenient and so the letters d, b and h are sometimes added to the end of a number to indicate the base. Thus, if we include upper and lower case characters, the quantity 19 decimal could also be represented as:

19   19d   19D   10011b 10011B 13h   13H

Microprocessor based systems tend to pass information around the system in groups of either eight or sixteen binary characters. It is important that the value of all characters is specified and so, if eight binary characters were being used, and we needed to represent the decimal value 7 it would be represented as 00000111. When hand writing binary quantities it is usual to separate them into groups of 4 characters and so the last number would become 0000 0111 or 0000 0000 0000 0111 if sixteen characters were being used.

G.2 CONVERTING BETWEEN BASES

For many microprocessor based applications you can specify numbers in either decimal, hexadecimal or binary and leave the software to do the conversions for you. However, when developing programs for a microprocessor based system to control external devices, there is a need for manual conversions of numbers with eight binary characters. This book therefore concentrates on the conversion of numbers of this size. In the unlikely event that you need to work with larger numbers, you should either extend the methods indicated here or consult another text for a more formal mathematical approach.

G.2.1 Binary to Hexadecimal Conversion

This is a very simple conversion to perform. The rules are:

    (a)   Start at the least significant digit and split the binary number into groups of four characters.

    (b)   Use Table G.1 and replace each group of four characters with its hexadecimal equivalent

e.g.1 $1011010_2 = 0101\ 1010_2 = 5A_{16}$ - (note the extra zero - 01011010)

e.g.2 $111010010000_2 = 1110\ 1001\ 0000_2 = E90_{16}$

G.2.2 Hexadecimal to Binary Conversion

This is an even simpler conversion. The rule is:

Replace each hexadecimal character by its four bit binary equivalent

e.g.1 $3_{16} = 0011_2$

e.g.2 $e39_{16} = 1110\ 0011\ 1001_2$

G.2.3 Binary to Decimal Conversion

If you examine Table G.1 you will notice that, starting from 1, each doubling of value adds a zero to the equivalent binary number :

decimal	binary	bit number
1	1	0
2	10	1
4	100	2
8	1000	3

For binary numbers with up to eight characters this can be summarized as

Table G.2

bit number	7	6	5	4	3	2	1	0
	128	64	32	16	8	4	2	1

To convert a binary number into decimal the rules are:

(a) Write the binary number underneath Table G.2 aligning the least significant character with bit number 0, the next least significant number with bit number 1 and so on.

(b) Add to together all the decimal numbers which are above a one to get the decimal equivalent.

e.g.1 1010_2

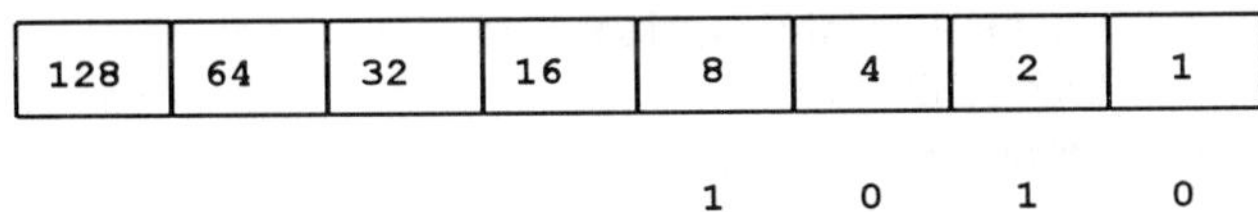

$$1010_2 = 8 + 2 = 10_{10} \qquad \text{(You can check this result with Table G.1)}$$

e.g.2 1101011_2

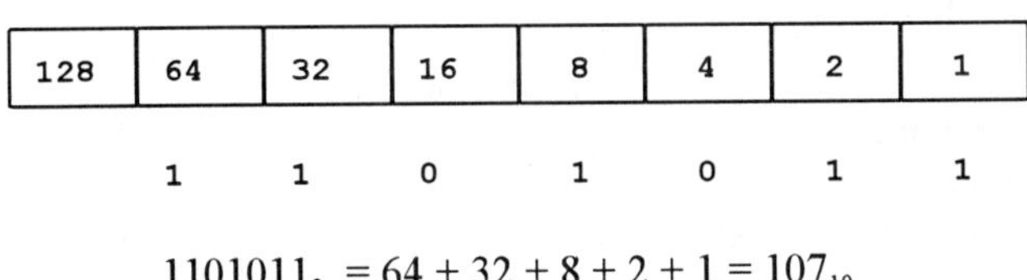

$$1101011_2 = 64 + 32 + 8 + 2 + 1 = 107_{10}$$

G.2.4 Decimal to Binary Conversion

In the following examples we are assuming that the decimal number to be converted is less than 256, If it is 256 or greater you will need to extend Table G.2. The procedure set out here is effectively reversing the procedure for binary to decimal conversion. As you work through you should regard the task as one of discovering where the ones were located in the original binary number. The rules are:

(a)   Find where the largest value was and subtract this from the number to be converted. Note the bit number and the result of the subtraction.

(b)   Treat the result of the subtraction as a new number and repeat (a).

(c)   Keep repeating (a) until you know the position of all the ones in the original binary number.

e.g.1 81_{10}

Bit 6 (64) is the largest that could have been used and so bit 6 must be a one.
$81 - 64 = 17$

Treating 17 as a new number bit 4 (16) must have been a one.
$17 - 16 = 1$

Treating 1 as a new number bit 0 (1) must have been a one.
$1 - 1 = 0$ and so we must have found all the ones in the original number.

We can now write the binary number putting ones at bits 6, 4 and 1 and zeros everywhere else:

$$81_{10} = 0101\ 0001_2\ = 101\ 0001_2$$

The procedure can be written more concisely as:

$$81 - 64 = 17 \qquad \text{(bit 6 = 1)}$$
$$17 - 16 = 1 \qquad \text{(bit 4 = 1)}$$
$$1 - 1 = 0 \qquad \text{(bit 0 = 1)}$$

$$\therefore \qquad 81_{10} = 101\ 0001_2$$

e.g.2 202_{10}

$$202 - 128 = 74 \qquad \text{(bit 7 = 1)}$$
$$74 - 64 = 10 \qquad \text{(bit 6 = 1)}$$
$$10 - 8 = 2 \qquad \text{(bit 3 = 1)}$$
$$2 - 2 = 0 \qquad \text{(bit 1 = 1)}$$

$$\therefore \qquad 202_{10} = 1100\ 1010_2$$

G.2.5 Hexadecimal/Decimal Conversion

Whilst a direct conversion is possible it is another set of rules which, unless used frequently, is easily forgotten . To limit the number of rules to be remembered it is suggested that conversions between decimal and hexadecimal are made via binary.

e.g.1 $6b_{16}$

$$6b_{16} = 0110\ 1011_2$$

Then, using binary to decimal conversion,
$$0110\ 1011_2 = 107_{10}$$

e.g.2 202_{10}

Using decimal to binary conversion
$$202_{10} = 1100\ 1010_2$$
and $\quad 1100\ 1010_2 = CA_{16}$

G.3 NEGATIVE NUMBERS

In microprocessor based equipment all information is passed around the system using only zeros and ones. This means that there is no character available to indicate if a value is positive or negative. The problem may be overcome by using the most significant character in a binary number to indicate the sign (zero indicates positive and one indicates negative). Thus, in an eight character binary number:

$$+6 \quad \text{may be represented as} \quad 0\ 000\ 0110$$
and $\quad -6 \quad$ *might* be represented as $\quad 1\ 000\ 0110$

The "might" for -6 has been emphasized since it is seldom, if ever, represented in exactly this format for applications involving the use of electronic equipment. For these applications negative numbers are usually represented in **2's complement** format. Again we shall omit the mathematical analysis and state the rule for expressing a binary number in 2's complement format as:

Invert all the characters and add one.

Thus if
$$+3 \quad = \quad 0000\ 0011_2$$

then, in 2's complement format,
$$-3 \quad = \quad 1111\ 1100\ +\ 1$$
$$= \quad 1111\ 1101$$

The value $1111\ 1101_2$ may not look like -3 but, when used for arithmetic operations with a set number of binary characters, it yields the correct result. Most microprocessors perform subtraction by adding the 2's complement and so, to evaluate 8 - 3 it will be treated as 8 + (-3). Thus, if we are working with eight binary characters we have:

$$
\begin{array}{rl}
8 & \texttt{0000 1000} \\
-3 & \texttt{1111 1101} \\
\hline
\end{array}
$$

when added gives   (1) `0000 0101` and $0000\ 0101_2$ = 5 - the correct result.

carry   1   1 1 1 1

Notice that bit 7 is a zero and indicates a positive result.

Similarly, for 3 - 5 we have

$5 = 0000\ 0101$ $\therefore$ $-5 = 1111\ 1010 + 1 = 1111\ 1011$

$$
\begin{array}{rl}
3 & \texttt{0000 0011} \\
-5 & \texttt{1111 1011} \\
\hline
 & \texttt{1111 1110}
\end{array}
$$

This may not look very much like the correct result but bit $7 = 1$ indicating a negative result. Also, we have been using the rules for 2's complement arithmetic and so a negative result will be in 2's complement format. To get back to our more usual format we must reverse the procedure for creating 2's complement

i.e.   subtract 1   1111 1110 becomes 1111 1101
        invert the characters   1111 1101 becomes 0000 0010
        and $0000\ 0010_2\ = 2$

Thus, remembering that it was a negative result, we have the correct answer of -2.

When working with eight binary characters we can represent 256 different quantities. If we are not using any convention to indicate the sign of the number then the smallest decimal value that can be represented is 0 and the largest is 255. If bit 7 is used to indicate the sign and we use 2's complement format the range is:

$$
\begin{array}{lll}
-128 & 1000\ 0000_2 & 80_{16} \\
-1 & 1111\ 1111_2 & ff_{16} \\
0 & 0000\ 0000_2 & 0_{16} \\
+1 & 0000\ 0001_2 & 1_{16} \\
+127 & 0111\ 1111_2 & 7f_{16}
\end{array}
$$

G.4 BINARY CODED DECIMAL

The numbering system which is frequently referred to as **Binary Coded Decimal (BCD)** by people involved in computing, is often called **Natural Binary Coded Decimal (NBCD)** by people involved in control engineering to distinguish it from other forms of binary coding such as Gray XS3 binary coded decimal. NBCD is often adopted when there is a need to display results in decimal format and consists of replacing each character in a decimal number by its **4-bit** binary equivalent.

e.g.1 295_{10}

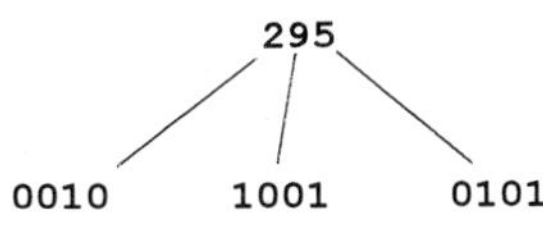

$$295_{10} = 0010\ 1001\ 0101_{NBCD}$$

e.g.2 $1000\ 0111\ 0110_{NBCD}$

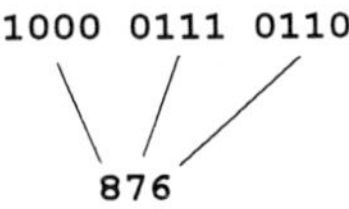

$$1000\ 0111\ 0110_{NBCD} = 876_{10}$$

Appendix H: I/O Board for Printer Port

This appendix gives details for adapting the printer port on a standard PC to provide general purpose input and output facilities.

Internally the printer port may be regarded as just three, TTL compatible, 8-bit ports at consecutive I/O addresses as shown below

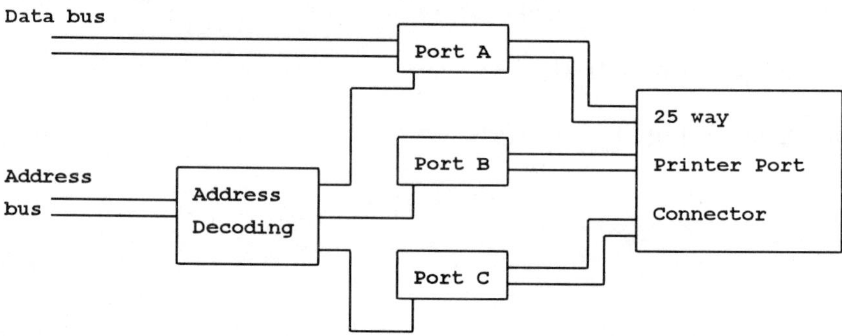

The addresses of these ports may vary from one PC to another but will be one of the following combinations:

port A	0x278	or	0x378	or	0x3bc
port B	0x279		0x379		0x3bd
port C	0x27a		0x37a		0x3be

Part of a PC's start up procedures involve establishing the address of the printer port and storing the address at memory location 00408 hex. If you examine the skeleton file given in Chapter 6 for 8086 programs you will see that it contains a routine for reading this address, calculating the address of port B and saving them both in labeled memory locations. This enables the student to refer to the parts of the printer port by name.

The ports are connected to a 25 way D type connector and, for printing, the pin functions are:

Pin	I/O	Signal	Port
1	O	Data Strobe	port C bit 0
2	O	Data bit 0	port A bit 0
3	O	Data bit 1	port A bit 1
4	O	Data bit 2	port A bit 2
5	O	Data bit 3	port A bit 3
6	O	Data bit 4	port A bit 4
7	O	Data bit 5	port A bit 5
8	O	Data bit 6	port A bit 6
9	O	Data bit 7	port A bit 7
10	I	Data acknowledge	port B bit 6 -active low
11	I	Busy	port B bit 7
12	I	Out of paper	port B bit 5
13	I	Selected	port B bit 4
14	O	Auto line feed	port C bit 1
15	I	Error status	port B bit 3
16	O	Initialize	port C bit 2
17	O	Select	port C bit 3
18	-	0 v	
19	-	0 v	
20	-	0 v	
21	-	0 v	
22	-	0 v	
23	-	0 v	
24	-	0 v	
25	-	0 v	

Whilst it is possible to connect directly to the printer port this is not advisable since there would be no protection between the expensive internals of the computer and any excessive signals generated by external equipment and/or careless student use. The circuit used by the author is shown in Figure H.2. It is not fool proof but does offer some measure of protection. If things go wrong it is more likely that the buffers (74LS04) will blow rather than the internals of the computer.

If you intend using this circuit with any inductive devices such as motors or relays you **MUST**, as a minimum, use suppression diodes and preferably use op-isolators.

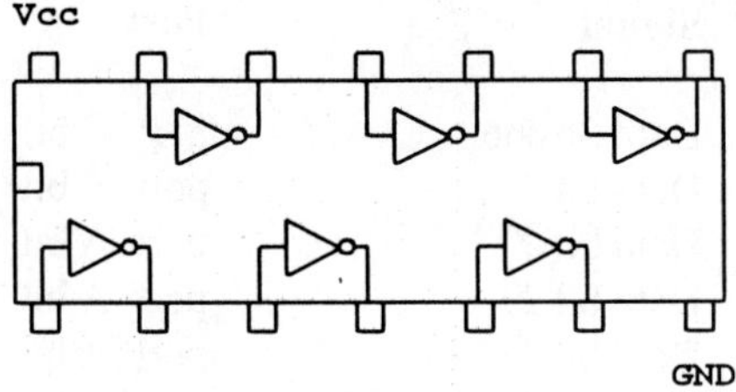

Figure H.1 - Pinout diagram for 74LS04

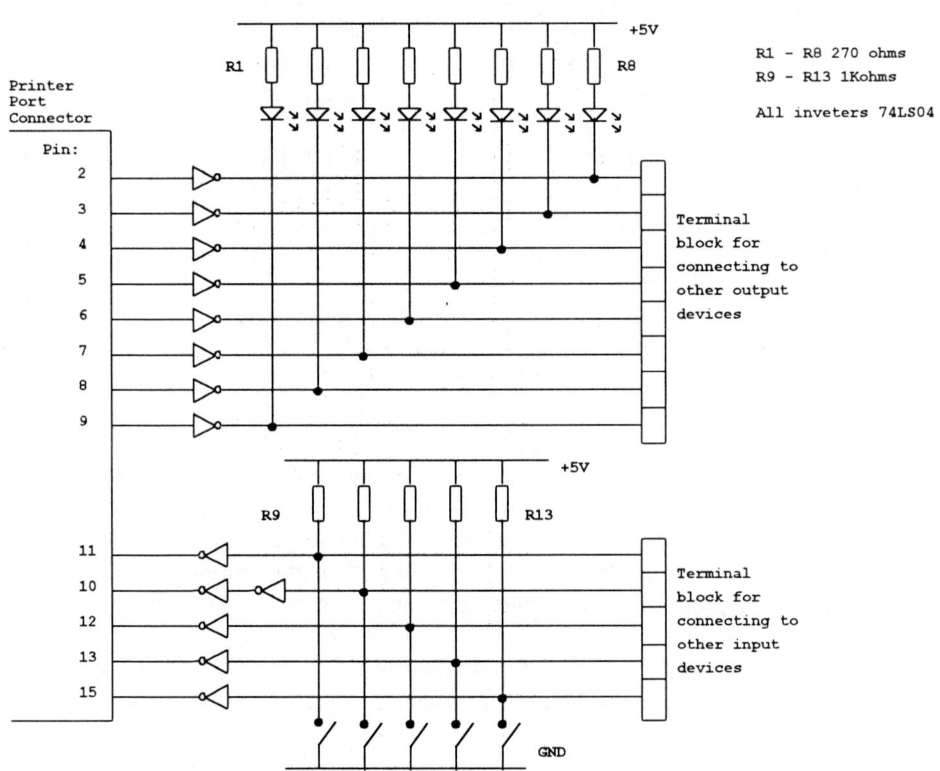

Figure H.2 Circuit for I/O Board

Answers to Exercises in Text

N.B. Some questions have been marked as "*** *suggested for portfolio* ***". Since these may be used for assessment, answers are not given. However, where appropriate, hints and suggestions as to a suitable approach have been included.

Exercise 1.2
The values established will depend upon the system being investigated. Differences between the maximum clocking frequency and the system clock are usually a design intention - the system has been designed to run at a lower frequency. The design supply voltage and the actual supply voltage should be fairly close - if they are not this is usually due to a faulty or inadequate power supply.

Exercise 1.5 *** *suggested for portfolio* ***
It is likely that you will have just started this course and you would not be expected to be able to provide an informed answer. Instead you would be expected to consider a possible application and use your experience with standard software packages such as word processors to identify likely routines. If you can not think of a good application have a look at the scenario for Case Study 3.

Exercise 2.1

	Z80	8086	68000
(a)	IR	no equivalent shown*	no equivalent shown*
(b)	PC	Instruction pointer	no equivalent shown*
(c)	BC, DE, HL	AX, BX, CX, DX	A0 to A7 for addresses D0 to D7 for data
(d)	ALU	ALU	ALU
(e)	Flags	Flags	CCR (condition code register)

* The reader should be aware that, although no equivalent is shown on the diagram, an equivalent must exist - it is the model which is incomplete, not the processor. When you search through manufacturers' data sheets you will often find that diagrams given as models are incomplete

Exercise 2.2
(a) The source operand is the accumulator shown as 'a'
   The destination operation is the address (1c34)
(b) Step 1 PC -> Address bus
       data at 1897 -> IR
       increment PC
    Step 2 PC -> Address bus
       data at 1898 -> MAR low
       increment PC

Step 3   PC -> Address bus
              data at 1899 -> MAR high
              increment PC
Step 4   MAR -> Address bus
              accumulator -> address 1c34

Exercise 2.3

(a)   1 op-code fetch, 3 memory reads, no memory writes

(b)   13

(c)   Your diagram should consists of one op-code fetch cycle as shown in Figure 2.4 followed by three memory read cycles as shown in Figure 2.2.

Exercise 2.5

Your answer should contain the following points
I/O Write:

1   address placed on address bus
2   data placed on data bus
3   IORQ and WR taken low indicating an I/O write operation and that the I/O device should read the data
6   IORQ and WR taken high indicating that I/O device should stop taking the data
7   processor stops putting the data on the data bus and the address is changed to the next operation

I/O read:

1   address placed on address bus
3   IORQ and RD taken low indicating an I/O read operation and that the I/O device should put its data on the data bus
4   processor starts to read the contents of the data bus
5   processor stops reading the data on the data bus
6   IORQ and RD taken high indicating that the I/O device should stop putting data on the data bus

Exercise 3.1   *** *suggested for portfolio* ***

Your answer should follow much the same sort of lines as that given in section 3.5. In each case there are several processors that might be chosen. You are expected to analyze the situations and select a processor according to *your* analysis.

Exercise 4.1

(a)   Pins marked I/O usually indicate input as well as output. Since data can be written to this device it is clearly a RAM.

(b)   There are 12 address lines therefore there are 2^{12} = 2048 memory locations. There are 8 data lines and so the organization is 2048 x 8 bit.
The size is 2048 x 8 = 16384 bits

Exercise 4.3   *** *suggested for portfolio* ***

The processor has only memory mapped I/O.   Also don't forget to consider the possibility of ghosting

Exercise 4.4

The main requirements are:

1.   the latches shall only be enabled when IORQ, WR and A0 through to A7 are all low.

2.   the tristate buffers shall only be enabled when IORQ, RD A1 through to A7 are all low and A1 is high.

Any logic circuit which achieves this may be used - one possibility is shown below:

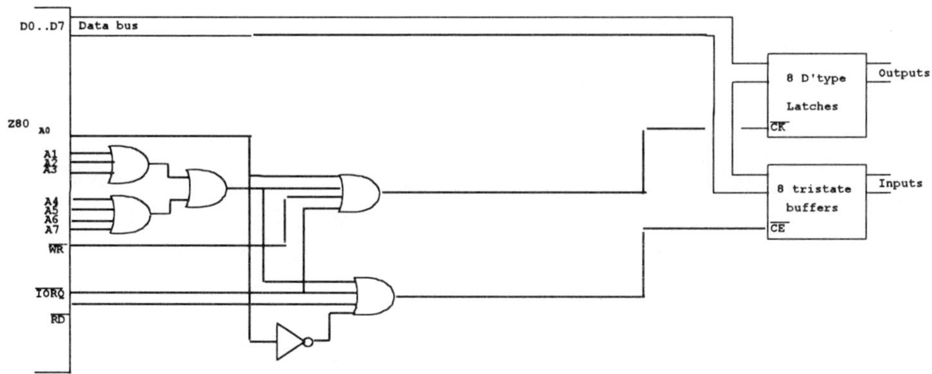

Exercise 5.1

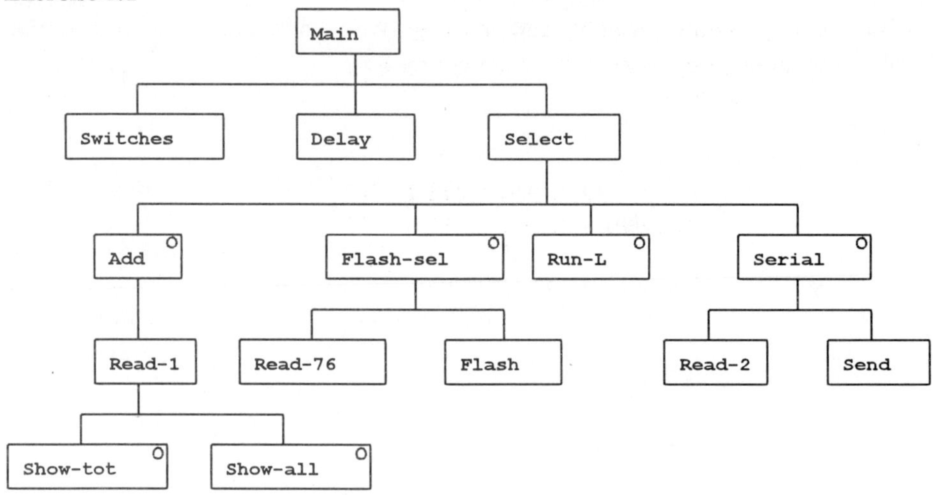

Data Structure Chart

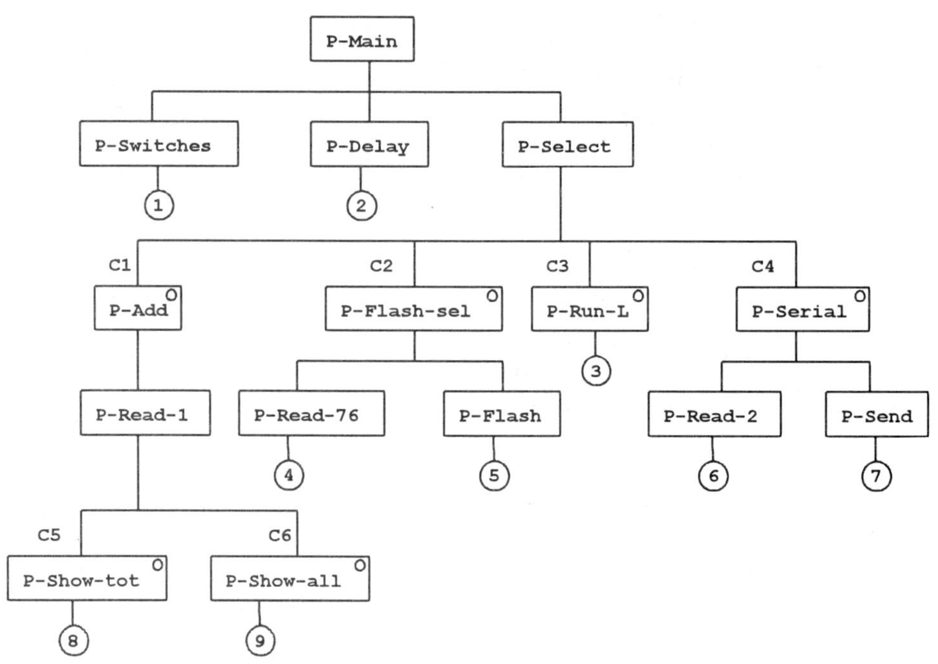

Program Structure Chart

Exercise 5.3

```
copy contents of address1234 in to register A
copy contents of address 1235 into register B
compare A with B
if A> B
{
        copy register A into address 1235
        copy register B into address 1234
}
end
```

Exercise 5.4

```
do
{
        read temperature sensor
        if temp < min
                turn heater on
        if temp > max
                turn heater off
}
```

Exercise 5.5

```
Main program P-Main
        read switches
        delay
        call P-Select

Subprogram P-Select
        if switches at 00
                call P-Add
        else if switches at 01
                call P-Flash-sel
        else if switches at 10
                call P-Run-L
        else
                call P-Serial

Subprogram P-Add
        call P-Read_1
```

```
Subprogram P-Read-1
        if total <= 99
                    display total on LEDs
        else
                    turn all LEDs on
```

The branches for conditions C2, C3 and C4 are expanded in a similar way.

Exercise 7.1

The following pseudo code is given for guidance:

```
get num1
get num2
add num1 and num2
get num3
add num3 to total of num1 and num2
store total at num4
```

Exercise 7. 2

Only the last line of the pseudo code needs to be changed to send the total to the LEDs instead of a memory location.

Exercise 7.4

The following pseudo code is given for guidance:

```
get num1
get num2
OR num1 with num2
get num3
AND num3 with the result  of num1 OR num2
display on LEDs
```

Exercise 7.5

The following pseudo code is given for guidance:

```
get num1
AND num1 with 0F₁₆
save result in a register - label res1
read switches
AND switches with F0₁₆ - label res2
OR res1 with res2
display result on LEDs
```

Exercise 8.1

	carry flag	overflow/ parity flag	zero flag	sign flag
(a)	clear	set	clear	set
(b)	set	clear	clear	clear
(c)	set	set	set	clear
(d)	clear	set	clear	set
(e)	clear	set	set	clear

Exercise 8.2

The following pseudo code is given for guidance:

```
get num1
get num2
compare num1 with num2   (num1 - num2)
if carry set
        display num1 on LEDs
else display num2 on LEDs
```

Exercise 8.3

The following pseudo code is given for guidance:

```
get num1
get num2
compare num1 with num2   (num1 - num2)
if z flag set
        illuminate all LEDs
else if carry set
        display num1 on LEDs
else display num2 on LEDs
```

Exercise 8.4

The following pseudo code is given for guidance:

```
set counter to 4
set pointer to first item of data
clear register for total

while counter not zero
{
        total = total + [pointer]
        increment pointer
        decrement counter
}
total = 0 - total
[pointer] = total
```

Exercise 8.5

The ASCII codes for the numbers 0 to 9 are 30_{16} to 39_{16} The following pseudo code is given for guidance:

```
set counter to 4
set pointer to first item of data

while counter not zero
{
        num = [pointer]
        num = num + 30₁₆
        send to printer
}
```

Exercise 8.6

When using a microprocessor to do arithmetic on data which is NBCD there is a risk that the result will not be NBCD e.g. $9_{NBCD} + 3_{NBCD}$ will be treated by the processor as $9_{16} + 3_{16}$ and give a result of C_{16} instead of 12_{NBCD}. Many processors, but by no means all, have instructions to help with this situation. The approach given in the code below will work for almost all processors.

```
read switches to register A
A = A + 55
copy A to register B

*** check lower nibble for valid NBCD  and adjust if needed ***
A = A AND 0F₁₆          - masks for lower nibble
A = A + 6               - if lower nibble > 9 result will be >= 10₁₆
if A >= 10
{
        A = B
        A = A + 6       - adjust lower nibble and add carry to upper nibble
        B = A
}
A = B
if A > 99
        illuminate all LEDs
else
        send A to LEDs
```

Exercise 10.1

The following pseudo code is given for guidance:

```
        clear register for total              (remember you may need to use 2 registers)
        set count to 8
        set pointer to start of data table


        while count not zero
        {
                total = total + [pointer]     (remember you may be using 2 registers)
                increment pointer
                decrement count

        }
        shift total 3 places right            (remember you may be using 2 registers)
        send result to LEDs
```

Exercise 10.2

The time delays required are

```
        4Hz       0.125s   (0.125s on and 0.125s off)
        2 Hz      0.25s
        1 Hz      0.5 s
        0.5 Hz    1s
```

Thus if we create a basic time delay of 0.125 seconds it can be executed the required number of times. The following pseudo code is given for guidance:

```
        Main program
                turn on LED 0
                do
                {
                        read switches into A
                        shift  A right 6 times      - moves bits 6 & 7 to bits 0 & 1
                        A = A + 1                   - number of times to repeat delay
                        copy to location labeled time_const
                        call delay
                        invert LED 0

                }

        Subroutine delay
                while time_const not zero
                {
                        execute 0.125s delay
                        decrement time_const

                }
```

Exercise 10.4

The following pseudo code is given for guidance:

```
set register A to 80₁₆
do
{
        send to LEDs
        call delay
        rotate A one place right
}
```

Exercise 10.5   *** suggested for portfolio ***

You should use the results of earlier work to establish a mark to space ratio which gives a suitable brightness for the simulated moon light. Dawn may then be simulated by increasing the mark to space ratio until the lights are full on and dusk by gradually changing it back again.

Exercise 11.1   *** suggested for portfolio ***

This is really just asking you to combine the information from section 11.1 with the pseudo code given in section 10.3 and translate it into a working program.

Exercise 11.3

A ramp of any sort may be achieved by steadily increasing (or decreasing) the digital input to a DAC. If we are using an 8-bit converter then, to get from minimum to maximum, in 1 second we will need to change the input every $1/256 = 3.92$ ms. Similarly to drop back to minimum in 0.1 s would require a delay of 0.392 ms. Thus the basic procedure becomes:

```
set count to zero
do
{
        do
        {
                send count to DAC
                delay 3.92 ms
                increment count
        }while count not  ff₁₆
        do
        {
                send count to DAC
                delay 0.392 ms
                decrement count
        }while count not 0
}
```

Index